PRACT
GOUR

*Company's
Coming*®

The
Canadian
Berry
Cookbook

Jeff Morrison
James Darcy

Library and Archives Canada Cataloguing in Publication
Morrison, Jeff, 1967-, author
        The Canadian berry cookbook / Jeff Morrison, James Darcy.

(Wild Canada series)
Includes index.
ISBN 978-1-988133-08-9 (wire-o)

        1. Cooking (Berries). 2. Cookbooks. I. Darcy, James, author
II. Title. III. Series: Morrison, Jeff, 1967- . Wild Canada.

TX813.B4M67 2016   641.6'47   C2014-908266-5

Distributed by
Canada Book Distributors - Booklogic
11414-119 Street
Edmonton. Alberta, Canada T5G 2X6
Tel: 1-800-661-9017

We acknowledge the financial support of the Government of Canada through the Canada Book Fund for our publishing activities.

Funded by the Government of Canada
Financé par le gouvernement du Canada | Canadä

PC: 30

Table of Contents

Introduction

Botanically speaking, berries are considered to be any fleshy fruit that is produced from a single ovary; all berries boast a thin covering and a fleshy interior. In layman's terms, berries are juicy, brightly coloured, sweet or sour and usually contain small seeds.

From a more personal and practical level, however, berries are wonderful, wild and wickedly versatile. Erupting with natural flavours and sugars, these sweet, magical jewels are a terrific source of vitamins, dietary fibre and natural antioxidants, and they are about as healthy a food as you will find anywhere. In addition to acting as the focal point in many of North America's favourite desserts, berries make the perfect accompaniment for many entrees and are the star of numerous side dishes. Berries make exceptional preserves, sauces, jams and jellies, and they also add refreshing flavour to cool beverages. When it comes to table fare, the beautiful berries of this country burst with flavours that are like an accent note in a melodic symphony.

Not only are berries delicious to eat, but they also come steeped in history and Canadian tradition. Native peoples used berries in a whole host of ways, from a primary food source to medicinal purposes, and even as a dye for clothing. Indigenous peoples discovered the merits of the glorious berry long ago, and their traditional practices serve as examples today of the global benefits to be gained by producing and consuming the various berries that grow in the provinces and territories of this country.

An almost ubiquitous natural resource, berries can be found in different forms and species throughout the hills and dales of Canada. From the Pacific coast to the northern forests to the Maritime bogs and meadows, big, bright, bold berries grow for only a short period each year. Thankfully, our farmers' markets and grocery stores carry some of the more popular varieties throughout the year.

In the *Canadian Berry Cookbook,* my goal is to enlighten and entertain while extolling the virtues of the wonderful berries we have at our disposal. There are so many terrific ways to include berries in our meals, and we will explore some of the finer and more intriguing aspects of cooking to perfection with this wondrous bounty. Along the way, we will delve into some of the peripherals regarding berries in Canada, such as berry management and foraging techniques, nutritional values, field identification, hazards and unsavoury berries, provincial practices, harvesting techniques and conservation. I will also share some of my own personal experiences with berries in my travels through the great outdoors of Canada. A virtual cornucopia of wild forage and natural foods can be found right under your nose. Now, off to the berry patch.

Foraging and Harvesting Wild Berries

The key to good, productive berry picking includes a few elements centred around how and when to harvest these sweet gems. Knowledge is power, so make it a point to educate yourself thoroughly on the berries you plan on picking. The most important element to harvesting berries, after correct identification of said berries, is the timing of your pick. Berries are meant to be harvested when they are ripe. The highest levels of natural sugars are found in ripe berries, making the ripe ones the most desirable. Ripe berries will give off what I refer to as a certain "prime for the picking" vibe. They can be easily plucked from the bush and are generally more robust and plump that their unripe neighbours. Understanding when a berry is ripe may take some practice but will eventually become instinctive.

When harvesting ripe berries, you should never remove more than about half the berries growing in any given area. A "half harvest" is a good rule of thumb to follow; doing so allows foraging for other pickers, as well as leaves some berries around for any wildlife in the area that rely on them. Leaving berries behind will require some self-control, but it is for the greater good.

You must be mentally and physically prepared when heading out to forage for berries. Some berry bushes carry thorns and are often found in locations that are less than desirable for humans—from steep mountainsides to soggy bogs—so dress accordingly, from your clothing to your choice of footwear. Long pants, long-sleeved shirts and solid boots are recommended. A long day of picking berries is a lot of fun but can be physical as well. A prepared berry picker is a happy berry picker.

Field Identification

As with any wild foods we harvest, a thorough knowledge of what we are picking is a must. Look at the photos on pages 159 and 160 to get a general idea of what the fruits look like in the wild. Get your hands on a copy of a reputable field guide; I recommend Lone Pine's Wild Berries series, which includes *Wild Berries of British Columbia, Wild Berries of Alberta, Saskatchewan and Manitoba,* and *Wild Berries of Ontario.* Read it, learn it and bring it along with you the next time you go afield in your quest for a pail full of wild goodness. Of course, those who prefer foraging for berries at the local market or grocery store will have the guess work taken out of the process, but other more adventurous types will require some education.

Berry Hazards

Perhaps the most important thing to keep in mind before trucking off to the nearest berry patch is that hazards do exist. The first obvious hazard is poisonous berries. The toxins in berries range from mild to powerful enough to kill you if eaten. Your field guide to the berries and other wild plants found in your area will go a long way to making sure poisonous berries do not end up in your pail.

There are other, less obvious hazards that exist while out berry picking. Because these tasty fruits appeal to wild animals, we must be aware that danger may lurk in and around the berry patch. During the summer months, grizzlies and black bears gorge themselves on wild berries to help build their fat reserves for winter. Always be mindful while picking in bear country that a bruin could show up at any moment. In some areas of higher bear density, I suggest carrying a can of bear spray as a deterrent. Better safe than sorry—nothing ruins a terrific day of foraging like being attacked by a marauding bear.

Bring bug spray too!

A Few Helpful Tips

Crêpes with Blueberry Compote

Serves 8

Crêpes can be filled and topped with a variety of delicious combinations, and blueberry compote is among the best—in my opinion. This recipe could have just as easily been called "Morrison Family Favourite." Perhaps it has something to do with having roots in the province of Québec, which helped foster our love for crêpes, but whatever the reason, my wife and I have passed that appreciation along to our children. Perfecting the crêpe does take time, as the batter is thinner and harder to work with than pancake batter. Using the right pans, finding the right batter thickness and knowing when to flip the crêpes all take practice, but the end result is well worth the effort.

2 cups (500 mL) fresh (or frozen, thawed) blueberries
1/4 cup (60 mL) granulated sugar
3 Tbsp (45 mL) water

1 cup (250 mL) flour
2 eggs
1/2 cup (125 mL) milk
1/2 cup (125 mL) water
2 Tbsp (30 mL) butter, melted
1/4 tsp (1 mL) salt

To make compote, place blueberries, sugar and first amount of water in blender, and pulse. Transfer mixture to small saucepan and heat on medium-low for 10 to 12 minutes, until thickened and it sticks to spoon. Set aside.

In large bowl, whisk together flour and eggs. Add milk, second amount of water, butter and salt, and beat until smooth. Preheat greased medium frying pan over medium-high. Pour or scoop batter into pan, approximately 1/4 cup (60 mL) for each crêpe. Tilt pan with a circular motion so batter coats surface evenly. Cook crêpe for 1 to 2 minutes, until bottom is light brown. Loosen with spatula, flip and cook other side for 1 to 2 minutes. Place crêpe on plate and spread some compote down middle, fold crêpe and serve immediately.

Wild Blueberry Pancakes with Blueberry Maple Syrup

Serves 4

Anyone who lives on the Canadian Shield has surely picked wild blueberries from time to time. There are so many productive wild blueberry regions in Canada, and so many recipes, including this one, for these wonderful berries. Some of my best wild blueberry picking took place near Peterborough in central Ontario. On my trips back to college in Lindsay at the end of summer, I would regularly park the old brown Plymouth along the side of Highway 7 so I could pick a basketful of blueberries. The bog country along that stretch of highway was the perfect habitat for blueberries, and I wasn't about to miss out.

> **4 Tbsp (60 mL) pure Québec maple syrup,** *divided*
> **2 cups (500 mL) fresh (or frozen, thawed) wild blueberries,**
>   *divided*
>
> **1 cup (250 mL) flour**
> **1 tsp (5 mL) baking powder**
> **1/2 tsp (2 mL) baking soda**
> **1 cup (250 mL) milk**
> **1 egg, beaten**
> **2 Tbsp (30 mL) melted butter**

In small saucepan, heat 2 Tbsp (30 mL) maple syrup until just warm. Add 1 cup (250 mL) blueberries to syrup and mash. Heat mixture just until it begins to boil. Remove from heat and set aside.

In large bowl, combine flour, baking powder and baking soda and mix well. In separate bowl, combine milk, egg, butter, remaining 2 Tbsp (30 mL) maple syrup and remaining 1 cup (250 mL) blueberries. Pour wet mixture into dry ingredients and mix well.

Preheat greased griddle to medium. Pour about 1/4 cup (60 mL) batter onto griddle. When deep holes or bubbles appear in pancake, flip and cook other side. Repeat for each pancake. Serve with blueberry syrup.

tip **PANCAKE SIZE**

Pancakes flip easily when they are made to match the size of your flipper.

Wild Blueberry Breakfast Bowl

Serves 2

Québec knows a thing or two about wild blueberries. The bulk of the province lies in the north, where wild blueberries grow *en masse*, as they say in French. Not only is there a huge recreational following for blueberries in La Belle Province, but the berries are big business as well. Since its creation in 1984, Québec Wild Blueberries Inc. (QWBI) is a company that has worked aggressively to develop products that exceed industry standards and are in tune with market demand. On top of participating in international food fairs, QWBI has invested time and energy into research and development to make them a commercial blueberry supplier to be reckoned with.

> 2 1/2 cups (625 mL) fresh or frozen wild blueberries
> 1 Tbsp (15 mL) brown sugar
> 1/2 tsp (2 mL) ground cinnamon
> 2 Tbsp (30 mL) ground flaxseeds
> 1/4 cup (60 mL) flaked coconut

In microwave-safe bowl, combine blueberries and brown sugar. Cook in microwave on high for 2 to 3 minutes, until mixture is hot, stirring once after 1 minute. Add cinnamon and stir. Add flaxseeds and coconut; stir until smooth. Serve in cereal bowls.

Healthy Blueberry Banana Muffins

Makes 12 muffins

The use of blueberry harvesting machines is growing in popularity. There are several styles of blueberry harvesters, and they all share some common similarities. The harvesters generally sport a set of picking heads that follow the contours of the ground and gently lift the branches and remove the berries. The picking heads work in conjunction with a comb-type picking bar, which in effect rakes the blueberries from the plants. The berries are deposited onto a conveyor belt that carries them to a capture bin. The advantage of a mechanical harvester is the ability to pick large quantities of berries in a short amount of time.

1 cup (250 mL) whole-wheat flour
3/4 cup (175 mL) all-purpose flour
1/4 cup (60 mL) wheat germ
1 tsp (5 mL) baking soda
1/2 tsp (2 mL) salt

2 ripe bananas
1/3 cup (75 mL) milk
1 tsp (5 mL) vanilla extract

1/2 cup (125 mL) butter, room temperature
1/3 cup (75 mL) granulated sugar
1/3 cup (75 mL) brown sugar
2 eggs

1 cup (250 mL) fresh or frozen blueberries

Preheat oven to 375°F (190°C). In medium bowl, mix together both flours, wheat germ, baking soda and salt.

In another bowl, mash bananas. Stir in milk and vanilla.

In large bowl, combine butter and both sugars and mix until light and fluffy. Add eggs, one at a time, mixing well after each addition. With mixer on low, alternately add flour mixture and banana mixture to butter mixture. Mix just until combined.

Fold in blueberries. Spoon batter into 12 greased muffin cups. Bake for 18 to 20 minutes, until wooden pick inserted in centre of muffin comes out clean.

Wild Blueberry Scones

Makes 16 scones

One of the last berries available for foraging each summer is the magnificent blueberry. A native of northern Canadian Shield forests, the blueberry is an iconic symbol of this country. Picking wild blueberries can be a fantastic day in the great outdoors, or it can be the worst day of back-breaking labour you've ever had—it's all in the approach. Picking any low-growing berry can cause back strain if you don't exhibit proper posture. Either crouching or sitting on a small stool are the preferred stances to ensure no pain at end of the day. Some areas may allow for sitting on the ground. Bending over to pick will cause back problems regardless of your age. Be comfortable, and try to pick with your hands straight out, not above your head and not below your knees.

4 cups (1 L) flour
6 Tbsp (90 mL) granulated sugar
4 1/2 tsp (22 mL) baking powder
1/2 tsp (2 mL) salt
1/2 cup (125 mL) plus 2 Tbsp (30 mL) butter

2 eggs
3/4 cup (175 mL) buttermilk

1 1/2 cups (375 mL) fresh or frozen wild blueberries
2 Tbsp (30 mL) milk

In large bowl, combine flour, sugar, baking powder and salt. Cut in butter until mixture resembles coarse crumbs.

In separate bowl, whisk together eggs and buttermilk. Add to dry ingredients and stir just until moistened.

Preheat oven to 375°F (190°C). Turn dough onto floured surface; gently knead in blueberries. Divide dough in half, and pat each half into 8 inch (20 cm) circle. Cut each circle into 8 wedges. Separate wedges and place on greased baking sheets. Brush with milk. Bake for 15 to 20 minutes, until tops are golden brown.

Chicken Salad with Blueberries

Serves 4

Black bears and grizzlies rely heavily on wild berries to maintain a nutritional balance and to keep fat reserves strong through the winter months. Bears spend much of their lives eating, and crops such as blueberries, saskatoon berries and gooseberries account for the bulk of their diet throughout the summer months. As with other naturally occurring plants, wild berry production varies from year to year. In years of scarcity, bears must change gears and supplement their diets with other foods such as insects, small mammals and even large mammals. Years when berry numbers are down are often years that human–bear encounters are at their highest.

3 Tbsp (45 mL) red wine vinegar
2 Tbsp (30 mL) raspberry jam (see Note)
1 Tbsp (15 mL) prepared mustard
1/4 cup (60 mL) olive oil
1/8 tsp (0.5 mL) salt
1/8 tsp (0.5 mL) pepper

1 x 4 1/2 oz (128 g) bag mixed salad greens
1 lb (454 g) cooked chicken breast, thinly sliced
1 small red onion, thinly sliced
8 oz (250 g) sugar snap peas, halved
1 cup (250 mL) fresh blueberries
1/4 cup (60 mL) slivered almonds, toasted

For dressing, combine vinegar, jam, mustard, oil, salt and pepper in small bowl. Mix well and set aside.

Layer lettuce, chicken, onion, peas and blueberries in large serving bowl. Pour dressing over salad, and toss. Sprinkle with slivered almonds. Serve immediately.

Note: If you don't want to use store-bought jam, you can make your own; Raspberry Honey Jam, page 79, will work well in this recipe.

Blueberry Tuna Salad

Serves 4

Bluefin tuna are one of the most commercially important fish species in the world. In Canada and the United States, bluefin tuna can only be caught on rod and reel, which is no easy task with a fish that can reach weights of over 1000 pounds (450 kilograms). With an Asian market paying upward of $20 per pound, commercial tuna fishing is big business—a single fish occasionally fetches $20,000 and more on the Asian market. Bluefin stocks have dwindled over past 50 years, but their numbers remain strong enough to support a commercial industry.

> 1 cup (250 mL) canned tuna, drained
> 1 cup (250 mL) diced celery
> 1 cup (250 mL) diced carrot
> 1 cup (250 mL) fresh blueberries
> 2 Tbsp (30 mL) mayonnaise
> 1 Tbsp (15 mL) green relish

Combine all 6 ingredients in medium bowl. Refrigerate for 1 hour before serving.

Blueberry Shrimp Salad

Serves 4

Isolating superior blueberries from substandard ones is not always easy. When picking your own blueberries, either in the wild or at a berry farm, try to focus on the berries that are at their peak. Stay away from those that are green or purple, as those ones are not yet ripe. Blueberries get their colour from high levels of anthocyanin, a pigment that ranges in colour from reddish to blue. A perfectly ripe blueberry, whether wild or cultivated, will be blue, firm and heavy for its size. Look for berries that have a white or silver surface bloom in their colouring. A perfectly ripe blueberry will also burst open with flavour when you bite into it.

1 tsp (5 mL) grated lemon zest
2 Tbsp (30 mL) lemon juice
1 Tbsp (15 mL) liquid honey
3 Tbsp (45 mL) olive oil
1/8 tsp (0.5 mL) salt
1/8 tsp (0.5 mL) pepper

1 x 8 oz (225 g) bag fresh spring mix
1/2 lb (225 g) cooked shrimp, peeled and deveined
1 cup (250 mL) fresh blueberries
1/2 cup (125 mL) feta cheese
1/3 cup (75 mL) pecan halves

In a small bowl, combine lemon zest, lemon juice and honey, and whisk well. Add olive oil, salt and pepper, and whisk again.

Arrange salad greens on 4 plates and top with shrimp, blueberries, feta and pecans. Drizzle with dressing and serve.

Pictured on page 33.

Braised Bison with Blueberry Marinade

Serves 6 to 8

Bison are raised in large ranches in the West. Although bison are range raised and fed, farmed in a similar way to our western beef cattle, that is where the comparisons ends. Bison meat is far superior to beef, in my opinion. It boasts a slightly sweeter taste, and most cuts will be more tender than beef. From a nutritional standpoint, bison far exceeds beef, being lower in cholesterol and lower in fat. Don't discount bison as a prime source of protein and a great alternative for the health conscious Canadian. The animals are raised in a controlled environment and except for being slightly more expensive than traditional red meat, they are perhaps the best beef alternative we have.

1/2 cup (125 mL) dry red wine
2 bay leaves
1/4 tsp (1 mL) ground allspice
salt and pepper
4 lbs (1.8 kg) bison roast

1/2 cup (125 mL) prepared beef broth
1 onion, sliced
4 bacon slices

1 cup (250 mL) blueberry juice

In small bowl, combine red wine, bay leaves, allspice, and salt and pepper to taste to form a marinade. Place roast in heavy-duty resealable plastic bag and pour marinade over roast. Seal bag and refrigerate for 24 hours.

Preheat oven to 350°F (175°C). Remove roast from bag, reserving marinade for later, and place in roasting pan. Add beef broth and onion to pan and place bacon on top of roast. Cook, uncovered, for 1 hour.

Mix together reserved marinade and blueberry juice. Pour over roast, cover and cook for 1 additional hour, until desired doneness.

Blueberry Turkey Meatballs

Serves 4

For a more natural twist, I sometimes use wild turkey meat in my Blueberry Turkey Meatballs recipe. The wild turkey, though native to this country, nearly disappeared in the late 19th and early 20th centuries owing to habitat loss and overhunting. Thanks to various restoration projects, where gobblers from the U.S. were released on Canadian soil, several provinces now have thriving wild turkey populations. The gobbler numbers in Ontario alone have topped 100,000, making the glorious wild turkey one of this country's greatest wildlife management success stories. I am very fortunate in my suburban Ottawa home to see wild turkeys in the back field on a regular basis.

> 3/4 cup (175 mL) fresh or frozen blueberries
> 2 tsp (10 mL) soy sauce
>
> 1 lb (454 g) ground turkey
> 1 Tbsp (15 mL) grated lemon zest
> 1 Tbsp (15 mL) minced ginger root
> 1 tsp (5 mL) pepper

Preheat oven to 350°F (175°C). Place blueberries and soy sauce in blender or food processor and process into sauce-like consistency.

Place ground turkey in large bowl. Add blueberry purée, lemon zest, ginger and pepper. Mix well and form into meatballs. Place meatballs on baking sheet. Cook for about 45 minutes, until well browned.

 tip **FREEZING BERRIES**

One simple technique for preserving those beautiful foraged berries for use at a later date is to freeze them. For the most part, freezing retains the nutrients and goodness of your berries. The important thing when freezing berries is to make sure they are completely dry. After rinsing the berries, place them in a strainer for 15 minutes and then spread them out on paper towel to dry. Arrange berries in a single layer on a rimmed baking sheet and place in freezer. Freezing them in a single layer keeps your berries from being frozen together in one big clump. Once frozen, remove from freezer and pour berries into an airtight container, then place back in freezer. They will keep for a year and more.

Almond-crusted Walleye with Blueberry Lime Butter

Serves 4

When preparing fillets for bony fish such as walleye, a good quality fillet knife is a must. The trick to using a filleting knife properly is to think of it as an extension of your hand. Hold the knife as if you were shaking someone's hand, and extend your pointer finger down the blade shaft for a more controlled cut. A precisely sharpened knife will cut cleanly and smoothly through the fish's flesh without being so sharp as to slice right through the smaller bones. With some practice, you will become proficient at filleting and deboning any fish you catch.

4 walleye fillets
salt and pepper
2 Tbsp (30 mL) vegetable oil, plus extra for brushing on fillets
1 cup (250 mL) bread crumbs
1 cup (250 mL) ground almonds, roasted
2 Tbsp (30 mL) chopped fresh parsley

1/2 cup (125 mL) butter, softened
1 tsp (5 mL) lime juice
1/2 cup (125 mL) fresh or frozen blueberries

Season fillets with salt and pepper, then brush them with a bit of oil. In medium bowl, combine bread crumbs, almonds and parsley. Dredge fillets through mixture to coat. Heat oil in frying pan over medium. Add fillets and cook for 5 to 6 minutes per side, until browned.

Place butter in bowl and beat until light. Add lime juice and blueberries. Mix well. Serve fillets topped with blueberry lime butter.

Grilled Trout with Wild Blueberry Sauce

Serves 4

Although trout and salmon are members of the same family, trout tend to be smaller and dwell in freshwater lakes and rivers, while most salmon are migratory, using streams and rivers to spawn but spending the rest of their lives in the ocean. Salmon tends to have a bolder taste than trout, and because the fish are larger, they are easier to fillet or have steaks cut. Smaller trout are often "cleaned" as opposed to filleted, but some members of the trout family such as lake trout and rainbow trout are good candidates for preparing boneless fillets. Trout flesh ranges from white to dark pink; as a rule of thumb, the larger the body of water they live in, the darker the flesh will be. Whether trout or salmon, bold or mild, blueberries are the perfect complement.

3/4 cup (175 mL) prepared chicken broth, *divided*
1/4 cup (60 mL) balsamic vinegar
1/4 cup (60 mL) orange juice
1 tsp (5 mL) honey
1 Tbsp (15 mL) cornstarch
1 cup (250 mL) fresh or frozen wild blueberries
2 tsp (10 mL) diced shallots

4 x 6 oz (170 g) trout fillets
2 Tbsp (30 mL) extra-virgin olive oil
1/2 tsp (2 mL) salt
1/2 tsp (2 mL) pepper

Combine 1/2 cup (125 mL) chicken broth, balsamic vinegar, orange juice and honey in saucepan. Bring to a boil over medium. Dissolve cornstarch in remaining 1/4 cup (60 mL) chicken broth and stir into simmering sauce. Cook, stirring, until sauce thickens and turns clear. Stir in blueberries and shallots, and reduce heat to low.

Preheat grill to medium. Brush trout with olive oil, and season with salt and pepper. Cook, turning often, for 8 to 10 minutes, until fish flakes easily with a fork. Serve with blueberry sauce.

Acorn Squash with Blueberries

Serves 4

Blueberries are big business. The blueberry industry of North America has been growing steadily as demand for the super berry grows. Canada is the world's second largest producer of blueberries both cultivated and wild, second only to the United States. According to Agriculture and Agri-food Canada, British Columbia leads the country in cultivated blueberry production, with over 13 million kilograms produced in 2010 alone. The leading wild blueberry producers in 2010 were Nova Scotia, New Brunswick and Prince Edward Island, which accounted for 95 percent of Canada's overall wild harvest that year.

2 acorn squash

1/4 cup (60 mL) brown sugar
3/4 tsp (4 mL) ground cinnamon
1 tsp (5 mL) cornstarch
1/4 tsp (1 mL) salt
1 cup (250 mL) fresh (or frozen, thawed) blueberries
3/4 cup (175 mL) diced apple
1 Tbsp (15 mL) butter

Preheat oven to 350°F (175°C). Cut squash in half and remove seeds. Place squash, cut side up, in baking dish filled with enough water to just cover bottom of dish.

In small bowl, combine brown sugar, cinnamon, cornstarch, salt and blueberries and apple. Spoon mixture into each squash cavity, then dot with butter. Cover and cook in oven for 45 minutes or until squash is tender. Spoon the squash out of the skin and into a bowl, and mash before serving.

Beautiful Blueberry Squares

Serves 12

Anytime you plan on foraging for your own berries, there is certain preparation and planning involved. Not only should you be prepared with the proper equipment and clothing, but your location needs to be researched as well. Many years ago, my wife and I had permission to go wild blueberry picking on the property of a natural science school just south of Mont-Tremblant. The foraging was incredible, and we already had a pail full of berries when a car pulled up along the small dirt road. The man was irate that we were picking near his property on a private road. We explained that we had permission to pick berries in that spot. He finally calmed down and apologized, but the incident exemplifies the need to secure permission whenever going onto or even crossing private property to forage.

3 Tbsp (45 mL) butter, softened
1/3 cup (75 mL) granulated sugar
2 egg yolks
2/3 cup (150 mL) flour
1 tsp (5 mL) baking powder
1/4 tsp (1 mL) salt
1/4 cup (60 mL) milk
1/4 tsp (1 mL) vanilla extract

2 egg whites
6 Tbsp (90 mL) granulated sugar
1/4 tsp (1 mL) salt
1 cup (250 mL) fresh (or frozen, thawed) blueberries

Preheat oven to 350°F (175°C). In medium bowl, combine butter and first amount of sugar. Mix. Beat in egg yolks. Add flour, baking powder and salt. Mix. Add milk and vanilla, and mix well. Transfer batter to 8 x 8 inch (20 x 20 cm) baking dish. Bake for 20 minutes.

In small bowl, beat egg whites until stiff. Beat in second amounts of sugar and salt. Fold in blueberries. Spread mixture over baked crust. Bake for another 15 to 20 minutes, until topping is slightly browned.

Pictured on page 34.

Moist Blueberry Cake

Serves 6

Blueberries are not generally regarded as large. A typical wild blueberry measures approximately 2.5 millimetres across. A typical cultivated berry is approximately 1 centimetre in diameter. In August 2008, a grower at Winterwood Farms in Kent, England, grew a blueberry about the size of a golf ball. The berry officially weighed in at 11.28 grams (0.4 ounces). Now that's a whole lot of blueberry, and at that size, only a dozen or so of the giant berries would be required for this entire cake recipe.

1 cup (250 mL) flour
1 tsp (5 mL) baking powder
1/2 tsp (2 tsp) salt

1/2 cup (125 mL) unsalted butter, softened
1 cup (250 mL) granulated sugar
1/4 tsp (1 mL) vanilla extract
2 eggs

2 cups (500 mL) fresh (or frozen, thawed) blueberries
1 tsp (5 mL) flour
1 tsp (5 mL) lemon juice
icing sugar for dusting

Preheat oven to 350°F (175°C). Combine first amount of flour with baking powder and salt. Set aside.

Using electric mixer, beat butter on medium-high speed for 2 minutes. Add granulated sugar and vanilla and beat until light and fluffy. Add eggs 1 at a time and beat until well blended. Reduce speed of mixer to low. Slowly add flour mixture and beat until smooth. Transfer batter to greased and parchment paper-lined 9 inch (23 cm) round baking pan.

In medium bowl, combine blueberries with second amount of flour and lemon juice. Spoon berry mixture over batter. Bake for 1 hour, or until wooden pick inserted in centre comes out clean. Remove cake from oven and allow to cool in pan for 10 minutes. Carefully slide a thin knife around edge of pan. Transfer cake to platter, berry side up. Dust with icing sugar before serving.

Blueberry Orange Cake with Cream Cheese Icing

Serves 12

According to the Ontario Berry Growers Association, blueberries are divided into two categories: highbush and lowbush. The wild blueberries growing in many parts of the province are the lowbush variety. Cultivated blueberries are generally the highbush variety; the shrubs grow taller and the berries grow larger, with the season running from early July to late August. Cultivated berries are available in a wide range of sizes; extra large usually account for 50 berries per 1 cup (250 mL), all the way down to small, which runs around 100 berries per cup.

3/4 cup (175 mL) butter
1 1/2 cups (375 mL) granulated sugar
3 eggs
1 tsp (5 mL) vanilla extract

2 cups (500 mL) flour
1/2 tsp (2 mL) baking powder
1/4 tsp (1 mL) salt
3/4 cup (175 mL) milk
2 cups (500 mL) fresh or frozen blueberries
grated zest from 2 oranges

4 oz (125 g) cream cheese, softened
1/4 cup (60 mL) butter, softened
2 cups (500 mL) icing sugar
1 tsp (5 mL) vanilla extract

Preheat oven to 350°F (175°C). In medium bowl, cream butter and gradually add granulated sugar, eggs and first amount of vanilla.

Combine flour, baking powder and salt and add to butter mixture, alternating with milk. Gently fold in blueberries and orange zest. Transfer batter to greased 10 inch (25 cm) tube pan. Bake for 50 to 55 minutes. Gently remove cake from pan and place cake on wire rack to cool.

For icing, combine cream cheese and butter and beat using electric mixer. Gradually add icing sugar and second amount of vanilla. If icing is too thick, add a few drops of milk. Ice cake and serve.

Blueberry Cherry Fudge

Makes 2 lbs (900 g) fudge

Although I was never a huge fan of fudge, there was always a certain *Je ne sais quoi* about nibbling on a few choice morsels. The problem I always found with fudge was that it was just too sweet for me. Then I discovered Blueberry Cherry Fudge. Introducing blueberry in conjunction with the cherry and vanilla flavouring creates a perfect marriage of taste and tones down the sweetness of traditional fudge. Give it a try and you will see what I mean.

> 1 1/2 lbs (680 g) white chocolate baking squares (24 squares), chopped
> 1 x 11 oz (300 mL) can sweetened condensed milk
> 1/8 tsp (0.5 mL) salt
>
> 1 tsp (5 mL) vanilla extract
> 1/3 cup (75 mL) dried blueberries
> 1/3 cup (75 mL) dried cherries

Combine chocolate, condensed milk and salt in medium saucepan over medium. Continue heating, stirring often, until chocolate has completely melted. Remove from heat.

Stir in vanilla, blueberries and cherries. Spread evenly in parchment paper-lined 8 x 8 inch (20 x 20 cm) baking pan. Chill for 2 hours, or until firm. Transfer to cutting board, peel off paper and cut into bite-size squares. Store in airtight container at room temperature.

No-bake Blueberry Strawberry Cheesecake Squares

Serves 6 to 8

Blueberries are synonymous with New Brunswick. They have grown wild in much of the province over the last 10,000 years and have now proven to be one of the fastest growth sectors in the province's economy, with revenues tripling over the last decade. New Brunswick accounts for approximately 12 percent of the world's production of wild blueberries, and development of the wild blueberry industry continues to serve as a model for economic growth. Not only are the berries delicious but they are also big business in New Brunswick.

1 1/2 cups (375 mL) graham cracker crumbs
1/4 cup (60 mL) butter, melted

4 x 8 oz (250 g) blocks cream cheese, softened
1/2 cup (125 mL) granulated sugar
1 cup (250 mL) fresh (or frozen, thawed) blueberries
1 cup (250 mL) stemmed and quartered fresh strawberries
2 cups (500 mL) frozen whipped topping, thawed

Mix graham crumbs and butter; press into bottom of 9 x 9 inch (23 x 23 cm) baking pan. Refrigerate.

In large bowl, beat cream cheese and sugar with electric mixer until well blended. Combine blueberries and strawberries in separate bowl and mash with fork. Stir berries into cream cheese mixture. Gently stir in whipped topping until swirled together. Spoon over crust. Refrigerate for at least 6 hours. Cut into squares and serve.

Blueberry Salsa

Makes about 3 cups (750 mL)

You might not automatically consider using berries in salty, savoury or spicy dishes, but they are more at home there than you may think. The natural sugars contained in berries provide an instant zing to the taste buds responsible for sweet tastes. Recent studies have shown that the taste buds are actually fooled into believing the food item is even sweeter than it is when accompanied by something salty. So next time you want to throw your taste buds in a whirl, try something sweet and salty. Or, in this case, sweet and spicy.

3 cups (750 mL) fresh blueberries, *divided*
1/4 cup (60 mL) lemon juice
3 Tbsp (45 mL) chopped fresh cilantro
2 jalapeño peppers, seeded and minced
1/3 cup (75 mL) diced red pepper
1/2 tsp (2 mL) salt

Chop (don't mash) 2 cups (500 mL) blueberries. In medium bowl, combine chopped berries with lemon juice, cilantro, jalapeño, red pepper, salt and remaining blueberries. Mix well. Use immediately, or cover and chill for 8 hours. Serve with fresh tortillas or tortilla chips. It will keep for 3 to 4 months in a sealed container in the refrigerator.

Blueberry

Blueberry Ginger Relish

Makes 1 1/4 cups (300 mL)

A recent study of middle-agers in Norway determined that eating 1 cup (250 mL) of berries each day for eight weeks increased their levels of HDL cholesterol, or good cholesterol as it is often called. Daily berry intake also lowered blood pressure in the patients included in the study. So there you go; our aging Canadian population would fare much better in their golden years with a daily dose of berries—not a very torturous medicine to swallow, and our health would be the ultimate winner.

> 1 cup (250 mL) coarsely chopped fresh blueberries
> 1 shallot, chopped
> 1 serrano chili, seeded and minced
> 1 Tbsp (15 mL) chopped fresh cilantro
> 1 Tbsp (15 mL) lime juice
> 1 tsp (5 mL) minced ginger root
> 1/4 tsp (1 mL) salt

Combine all 7 ingredients in small bowl and mix well. Refrigerate for minimum 30 minutes. Serve with grilled pork or chicken. It will keep for up to 6 months in a sealed container in the refrigerator.

Blueberry Jelly

Makes 12 cups (3 L)

Blueberry jelly has a very sentimental attachment for me to my youth. One morning before heading off for a day of fishing on nearby Beaven Lake, my friend Patrick invited me in for a quick breakfast: toast with blueberry jelly. Blueberry jelly was not something that my own parents made, so it was a real treat for me. Although over 35 years have passed since then, I have never forgotten my first taste of blueberry jelly. My wife and I now make it regularly, and it brings me back to that day of fishing and blueberry jelly on toast every time. Even though no fish were caught on that outing, it just didn't seem to matter.

8 cups (2 L) fresh blueberries
4 cups (1 L) water

12 cups (3 L) granulated sugar
2 x 3 oz (85 mL) pouches liquid pectin

Crush blueberries slightly and then place them in Dutch oven. Add water. Bring to a boil over medium-high. Reduce heat to medium and cook, uncovered, for 45 minutes. Line strainer with 4 layers of cheesecloth and place over large bowl. Pour blueberry mixture into strainer; cover with edges of cheesecloth. Let stand for 30 minutes, or until liquid measures 6 cups (1.5 L).

Pour juice back into Dutch oven; gradually stir in sugar until it dissolves. Bring to a boil over high, stirring constantly. Add pectin; bring to a full rolling boil for 1 minute, stirring constantly. Remove from heat. Skim off any foam. Carefully ladle hot mixture into six 2 cup (500 mL) hot sterile jars, leaving 1/4 inch (6 mm) headspace. Wipe rims clean. Place hot metal lids on jars and screw on metal bands fingertip tight. Process in boiling water bath for 15 minutes. Remove jars and allow to cool. Check that jars have sealed properly. Store sealed jars in a cool, dark place.

Blueberry

Electric Blueberry Martinis

Makes 20 martinis

The martini is one classy drink indeed. However, the classic martini has become a slight bit humdrum. Enter the beloved blueberry, which turns your run-of-the-mill martini into an electrifying, colourful treat. Not only will the appearance of your martini change drastically when you add blueberry, producing a vibrant blue sheen, but there will also be an interesting freshness and sweetness that you won't get in the typical martini. It is a great summertime twist on an already proven beverage.

> 1 x 32 oz (1 L) bottle vodka
> 1 cup (250 mL) fresh blueberries
>
> 2 cups (500 mL) raspberry liqueur
> 5 limes, juiced
> 20 twists of lime zest, as garnish

Pour out 1 1/3 cups vodka (325 mL) into holding container; set aside. Using a wooden pick, poke a hole in each blueberry and then add blueberries to vodka bottle. Fill vodka bottle with vodka previously set aside (you'll have some leftover). Place bottle in a cool, dark place for 2 weeks.

To make martinis, combine 2 parts blueberry vodka, 1 part raspberry liqueur, and a dash of lime juice with ice in martini shaker. Shake vigorously and strain into glasses. Garnish each one with a twist of lime zest.

Cranberry Porridge

Serves 4

Like their blueberry cousins, cranberries in Canada can be found both wild and domesticated. Wild cranberries are generally smaller but are packed with a flavour that seems disproportionate to their size. Although they can be very tart, they make a refreshing trail nibble. And because they remain on the shrubs all year, wild cranberries can be a valuable survival food, rich in vitamin C and antioxidants. They are low-growing and can be difficult to collect; however, they are worth the effort.

> 2 cups (500 mL) milk
> 1 1/4 cups (300 mL) water
> 1/4 tsp (1 mL) ground cinnamon
> 1 1/3 cups (325 mL) rolled oats
> 1/3 cup (75 mL) dried cranberries
>
> 1/4 cup (60 mL) slivered almonds, roasted

In saucepan, combine milk, water and cinnamon. Bring to a boil over medium-high, stirring often. Gradually add rolled oats and dried cranberries. Reduce heat and let simmer, stirring often, for 3 to 5 minutes, until oatmeal is tender.

Divide into 4 bowls, and top with almonds. Serve immediately.

 DRYING BERRIES

Drying berries is an excellent way to preserve them if you have more than you can use fresh. An easy way is to spread them out on a screen and place it in a sunny spot outdoors. You may want to place protective netting over the berries to protect them from foragers. As the berries dry out, you will notice the size and shape change over time. Roll the berries around on the screen occasionally to expose the entire berry to direct sunlight. Be sure to bring them in at night to keep the dew off. The berries may take two or three sunny days to fully dehydrate.

Cranberry Oatmeal Breakfast Cookies

Makes 36 cookies

Cranberries are big business in British Columbia. The BC Cranberry Growers' Association represents approximately 80 growers that produce up to 40 million kilograms of wonderful cranberries each year. Most of the berries are destined for the juice market, but there is also demand for cranberry wines as well as dried, fresh and frozen berries. Farm sales for 2005 were valued at $34 million. The province is currently the largest cranberry producer in Canada and makes up approximately 12 percent of North America's entire cranberry production.

1 cup (250 mL) rolled oats
2/3 cup (150 mL) flour
1/2 cup (125 mL) wheat germ
1 tsp (5 mL) baking powder
1/2 tsp (2 mL) salt

1/2 cup (125 mL) butter, softened
1/2 cup (125 mL) brown sugar
1 egg
1/2 tsp (2 mL) vanilla extract

3/4 cup (175 mL) dried cranberries
1/4 cup (60 mL) chopped pecans

Preheat oven to 375°F (190°C). In medium bowl, combine rolled oats, flour, wheat germ, baking powder and salt. Mix.

In large bowl, cream together butter and brown sugar. Add egg and vanilla. Beat until smooth. Gradually add oat mixture. Beat on low until well mixed.

Mix in cranberries and pecans. Drop dough, using 1 Tbsp (15 mL) for each cookie, about 1 inch (2.5 cm) apart onto 2 greased baking sheets. Flatten slightly. Bake on separate racks in oven for about 10 minutes, switching position of baking sheets at halftime, until golden. Transfer cookies to wire racks to cool.

Cranberry Chokecherry Dip

Serves 6 to 8 as an appetizer

So you have too many cranberries on hand and would like to save them for use sometime in the future? Storing cranberries, and most other berries, is actually quite a simple task. Be sure to separate the substandard berries from the rest before refrigerating or freezing them. Cranberries will last one month or more in an airtight container in the fridge, provided that they are dry when first stored. If you freeze them in a plastic freezer bag, with the air removed, they can last up to one year and, when thawed, will taste as fresh as the day you picked them.

> 2 cups (500 mL) fresh (or frozen, thawed) cranberries
> 1 cup (250 mL) granulated sugar
>
> 1 cup (250 mL) chokecherry jam (see Note)
> 1 cup (250 mL) chopped pecans
>
> 1 x 8 oz (250 g) block cream cheese

Preheat oven to 350°F (175°C). Combine cranberries with sugar in medium baking dish with lid, stirring well. Bake, covered, for about 30 minutes, until cranberries pop and release their liquid.

Stir in chokecherry jam and pecans. Refrigerate for at least 2 hours.

To serve, allow cream cheese to come to room temperature, and pour cooled cranberry mixture over cream cheese. Serve with crackers of your choice.

Note: If you don't want to use store-bought jam, you can make your own; Traditional Canadian Chokecherry Jam, page 139, will work well in this recipe.

Blueberry Shrimp Salad (p. 15),
Blackberry Lime Soda (p. 82)

Beautiful Blueberry Squares (p. 21)

Cranberry Camembert Pizza

Serves 6 to 8 as an appetizer

This pizza appetizer features the popular cheese Camembert, but it could have just as easily called for brie; both are soft, white cheeses that originate in France. Brie was originally made in southeastern of France, while Camembert originated in the northwest on the coast of Normandy, where a different breed of cattle grazed on green pastures. Brie is typically formed in a large, flat wheel, and Camembert in a small cylinder. Although similar in many ways, they are also quite different. Brie tends to be mild, while Camembert has a bolder taste and a more pungent smell.

1 x 8 1/2 oz (235 g) tube refrigerator crescent rolls
8 oz (250 g) Camembert cheese, crumbled
3/4 cup (175 mL) cranberry sauce (see Note)
1/2 cup (125 mL) finely chopped pecans

Preheat oven to 400°F (200°C). Unroll crescent roll triangles and spread evenly in lightly greased 12 inch (30 cm) pizza pan; gently press together. Bake for 5 to 7 minutes, until lightly browned. Remove from oven and sprinkle with Camembert. Spoon cranberry sauce over cheese and then top with pecans. Place back in oven for another 5 minutes, until cheese is melted and crust is golden brown. Allow to cool for 5 minutes, then cut into wedges and place on serving platter.

Note: If you don't want to use store-bought sauce, you can make your own; Whole-berry Cranberry Sauce, page 46, will work well in this recipe.

Cranberry Microwave Party Meatballs

Serves 10 as an appetizer

The next time you put together a batch of Cranberry Microwave Party Meatballs, consider using ground venison instead of ground beef. The earthy taste and bold flavour of the venison will add an extra element to your meatballs and spruce up any party. Venison is typically leaner and boasts a more gamey flavour; however, when combined with cranberry sauce and the other ingredients in this recipe, the gaminess will be toned down. Use wild game meat anytime ground beef is called for—you may be surprised by how much you like it.

1 1/2 lbs (680 g) ground beef (or venison)
2 eggs
1 x 1 1/4 oz (38 g) envelope onion soup mix
1/2 cup (125 mL) bread crumbs

1 1/2 cups (375 mL) cranberry sauce (see Note)
3/4 cup (175 mL) ketchup
1/2 cup (125 mL) prepared beef broth
3 Tbsp (45 mL) diced onion
3 Tbsp (45 mL) brown sugar
2 tsp (10 mL) chili sauce

20–30 cocktail picks

Combine beef, eggs, onion soup mix and bread crumbs in large bowl. Mix well. Fashion into 1 inch (2.5 cm) balls. Place half of meatballs on microwave-safe plate and cover with waxed paper. Microwave on high for 4 to 5 minutes, until meat is no longer pink. Place cooked meatballs on paper towel to drain. Repeat with remaining meatballs.

In separate microwave-safe dish, combine cranberry sauce, ketchup, beef broth, onion, brown sugar and chili sauce. Cover with waxed paper and microwave on high for 4 minutes, stirring once halfway.

Place meatballs in serving tray, pour sauce over top and insert 1 cocktail pick in each meatball. Serve warm.

Note: If you don't want to use store-bought sauce, you can make your own; Whole-berry Cranberry Sauce, page 46, will work well in this recipe.

Cranberry Spinach Salad

Serves 4

We all know that cranberries are delicious, but did you know they're also good for preventing urinary tract and kidney infections? Research has proven that cranberries and cranberry juice will go a long way in eliminating urinary tract infections and keeping your kidneys functioning properly. The inherent acidity of the cranberry keeps bacteria from attaching to the walls of the bladder, which means that infections cannot stick around inside the body. So go ahead and enjoy your Cranberry Spinach Salad.

2 cups (500 mL) baby spinach leaves
1 cup (250 mL) dried cranberries
4 oz (125 g) goat cheese, crumbled
1/2 cup (125 mL) slivered almonds

2 Tbsp (30 mL) balsamic vinegar
1 Tbsp (15 mL) honey
1 tsp (5 mL) Dijon mustard
1/4 tsp (1 mL) pepper
1/4 cup (60 mL) extra-virgin olive oil

In large salad bowl, toss baby spinach, cranberries, goat cheese and almonds.

Mix balsamic vinegar, honey, Dijon mustard and pepper with whisk until well blended. Gradually add oil, whisking constantly until well blended. Pour over salad; toss to coat. Serve immediately.

 EXTRA-VIRGIN OLIVE OIL

You may have noticed that many recipes call for extra-virgin olive oil, and you may have wondered—why? When olive oil is made, the most pure variety and that which is intended for human consumption is referred to as "extra virgin." This high-quality oil has a wonderful aroma and adds a subtle sweetness to foods during the cooking process. Extra-virgin olive oil contains antioxidants and is great from a health standpoint. It costs a bit more than regular oil but is worth the expense.

Slow Cooker Cranberry Roast

Serves 8

Upper Canada Cranberries, an Ottawa-area farm, is one of only three commercial cranberry producers in Ontario. Lyle Slater started his venture in 1996 after reading an article on cranberries and realizing his farm land was perfect for growing the berries; he had the highly acidic soil and peat moss of a former bog, and a ready water source in the creek that ran through his property. In 2008, Upper Canada Cranberries expanded to include a small line of cranberry-based products. Look for them if you're ever in the area.

3 lbs (1.4 kg) beef chuck roast
2 cups (500 mL) cranberry sauce (see Note)
1 x 1 1/4 oz (38 g) envelope onion soup mix

2 Tbsp (30 mL) butter, melted
2 Tbsp (30 mL) flour

Place roast in slow cooker. Top with cranberry sauce and sprinkle with onion soup mix. Cover and cook on Low for 10 hours. Remove roast and set aside.

Turn slow cooker to High. Whisk butter and flour together and mix into liquid remaining in slow cooker to create gravy. Serve with roast.

Note: If you don't want to use store-bought sauce, you can make your own; Whole-berry Cranberry Sauce, page 46, will work well in this recipe.

Cranberry Chicken

Serves 8

When preparing a delicious dish such as this one, the cranberries aren't the only item you need to take special care with. Any recipe that calls for chicken must be handled with extreme caution because raw chicken can carry unwanted bacteria. Be sure when cutting raw chicken into serving-size pieces that your cutting surface is thoroughly cleaned afterward, and keep your cranberries away from the raw meat and cutting surface at all times. Bacteria from raw meat can contaminate the rest of your ingredients; if you're not careful, you could turn an otherwise superb meal into your worst nightmare.

1/4 cup (60 mL) flour
1/2 tsp (2 mL) salt
8 boneless, skinless chicken breast halves, cut into
  bite-sized pieces

1/4 cup (60 mL) vegetable oil

1 1/2 cups (375 mL) fresh cranberries
1/2 cup (125 mL) granulated sugar
1/2 cup (125 mL) orange juice
1 Tbsp (15 mL) grated orange zest
1/4 tsp (1 mL) ground ginger

Combine flour and salt in resealable plastic bag. Add chicken and toss until coated.

Heat oil in cast-iron skillet over medium. Add chicken pieces to skillet and cook just until beginning to brown, being careful not to cook the chicken fully.

Combine remaining 5 ingredients in saucepan over medium-high. Bring to a boil and pour over chicken in skillet. Cover skillet, reduce heat and simmer for 30 to 40 minutes, until chicken is tender.

Festive Chicken Bake

Serves 6

Festive Chicken Bake cries out to be eaten with family and friends. It is a true comfort food best enjoyed during the cold winter months when relatives stop by from out of town. I like to serve this dish on Boxing Day or around New Year's because, for a lack of a better description, it's a meal that sticks to your ribs and feels good in your tummy. The hint of cranberry and orange will remind you of warm summer days not so far off.

3 Tbsp (45 mL) flour
1/4 tsp (1 mL) salt
1/8 tsp (0.5 mL) pepper
1/8 tsp (0.5 mL) paprika
6 boneless, skinless chicken breast halves

1 Tbsp (15 mL) vegetable oil

1/2 cup (125 mL) prepared chicken broth
1 cup (250 mL) fresh or frozen cranberries
1/3 cup (75 mL) frozen concentrated orange juice, thawed
1/4 cup (60 mL) diced onion
1/4 tsp (1 mL) ground cinnamon
1/4 tsp (1 mL) ground ginger

Combine flour, salt, pepper and paprika in large resealable plastic bag. Add chicken and toss until coated. Remove chicken. Reserve remaining flour mixture.

Heat oil in large frying pan on medium-high. Add chicken. Cook for 2 to 3 minutes per side until browned. Arrange chicken in single layer in greased 2 quart (2 L) shallow baking dish.

Preheat oven to 350°F (175°C). In small bowl, combine chicken broth and reserved flour mixture. Mix. Stir in remaining 5 ingredients. Pour over chicken. Bake, covered, for about 45 minutes, until fully cooked.

Pictured on page 51.

Slow Cooker Turkey Breast with Cranberry Sauce

Serves 6 to 8

There are two ways to acquire a whole turkey breast. One is to remove the breast from a thawed whole turkey, which can be a bit of a chore depending on your knife skills. A simpler way is to purchase a boneless turkey breast from the grocery store. These are whole prepared boneless turkey breasts and are sometimes pre-stuffed, depending on the configuration and manufacturer. For this recipe, look for whole boneless turkey breasts that are not stuffed because they work better in the slow cooker. Be sure to thaw the meat prior to starting your slow cooker. Use the rule of thumb of one hour of thawing time per pound.

2 cups (500 mL) cranberry sauce (see Note)
1 x 1 1/4 oz (38 g) envelope onion soup mix
1/2 cup (125 mL) orange juice

butter, for greasing slow cooker
2 to 3 lbs (900 g to 1.4 kg) boneless turkey breast

In medium bowl, mash cranberry sauce. Add onion soup mix and orange juice. Mix well. Set aside.

Grease inside of slow cooker with butter. Add turkey breast. Pour sauce over turkey. Cook on Low for 6 to 7 hours, until turkey is fully cooked. Serve turkey with sauce.

Note: If you don't want to use store-bought sauce, you can make your own; Whole-berry Cranberry Sauce, page 46, will work well in this recipe.

Cabbage with Cranberries

Serves 6

Cranberries are a highly versatile berry and can be used in a wide variety of sweet and savoury recipes, and with a few tricks, they can be tweaked and modified for uses you wouldn't commonly think of. One trick that some chefs use is to neutralize the acid of cranberries by adding a touch of baking soda, which facilitates the use of less sugar. In addition to being a little sour, cranberries can be tough. Heating them until they pop significantly softens them, a fact that many recipes take into account in the cooking time. And if a recipe calls for fresh cranberries, dried cranberries can often be substituted with pleasing results.

1 Tbsp (15 mL) butter
1/2 small red onion, chopped

1/2 head cabbage, shredded

1 cup (250 mL) prepared chicken or vegetable broth
1 cup (250 mL) cranberry juice
1/4 cup (60 mL) fresh or dried cranberries
1 Tbsp (15 mL) red wine vinegar
1/4 tsp (1 mL) salt
1/4 tsp (1 mL) pepper
1/8 tsp (0.5 mL) ground cinnamon
1 bay leaf

Melt butter in large skillet over medium. Add onion and cook for about 5 minutes, until softened.

Add cabbage and cook, stirring, for about 2 minutes, until coated.

Stir in broth, cranberry juice, cranberries, vinegar, salt, pepper, cinnamon and bay leaf, and bring to a boil. Reduce heat to low; cover and simmer for about 45 minutes, until cabbage is almost tender. Uncover and simmer, stirring occasionally, for about 20 minutes, until cabbage is tender and most of the liquid is absorbed. Discard bay leaf before serving.

Thanksgiving Cranberry Stuffing

Serves 10

To make an even more traditional Thanksgiving dinner with your cranberry stuffing, why not try wild turkey? Wild gobbler is one of the least gamey-tasting meats around and could easily be mistaken for commercial or farm-raised turkey. Whether you tell your guests it is the wild variety is strictly up to you. My wife and I always like to be honest with people in advance in case anyone has an aversion to wild game. It also gives me an opportunity to discuss the merits of game meat and also the finer points of responsible hunting and conservation.

> 1 cup (250 mL) butter
> 3 celery ribs (with leaves), chopped
> 3/4 cup (175 mL) diced onion
>
> 1/2 cup (125 mL) dried cranberries
> 1 1/2 tsp (7 mL) dried sage
> 1 tsp (5 mL) dried thyme
> 1 1/2 tsp (7 mL) salt
> 1/2 tsp (2 mL) pepper
>
> 15 slices bread, cut in cubes

Melt butter in a large skillet over medium. Add celery and onion and cook, stirring frequently, until onion is tender. Remove from heat.

Add cranberries, sage, thyme, salt and pepper. Stir to combine.

Place bread cubes in a large bowl, add contents of skillet and toss. Stuff turkey prior to cooking (see Note).

Note: If you prefer not to stuff the turkey, you can simply place the ingredients in a greased 3 quart (3 L) casserole. Cover and bake in a 325°F (160°C) oven for 30 minutes. Uncover and bake for 15 minutes longer.

Berry Full Wafers

Makes 36 cookies

The Ontario Berry Growers Association (OBGA) is one of many provincial organizations dedicated to representing berry growers, berry pickers and all related facets of the berry industry. The OBGA is made up of approximately 200 volunteer growers, who account for 80 percent of the berry crop in Ontario. The goals of the association are to foster the advancement of the berry industry in Ontario, to promote better harvesting and marketing practices, and to encourage and promote all activities related to the Ontario berry industry. Fresh berries can be enjoyed from May to October, and the OBGA makes it their mission to ensure that that continues.

1/2 cup (125 mL) finely chopped dried cranberries
1/2 cup (125 mL) icing sugar
1/2 cup (125 mL) spreadable strawberry cream cheese
1 tsp (5 mL) raspberry liqueur, optional

72 vanilla wafers

12 x 1 oz (28 g) semi-sweet chocolate baking squares, chopped
1 Tbsp (15 mL) butter

1 x 1 oz (28 g) white chocolate baking square, chopped

Combine first 4 ingredients in medium bowl to make berry filling.

On each of 36 wafers, spread 1 tsp (5 mL) filling. Place remaining 36 wafers on top of filling. Chill for about 1 hour, until filling is firm.

Heat semi-sweet chocolate and butter in small heavy saucepan on low, stirring often until chocolate is almost melted. Remove from heat. Stir until smooth. Place 1 cookie on top of fork. Dip into chocolate mixture to completely coat, allowing excess to drip back into pan. Place on waxed paper-lined baking sheet. Repeat with remaining cookies and chocolate mixture. Let stand until set. May be chilled to speed setting.

Heat white chocolate in separate small heavy saucepan on low, stirring often until almost melted. Remove from heat. Stir until smooth. Drizzle white chocolate in decorative pattern over each dipped cookie. Let stand until set.

Cranberry

Maritime Cranberry Pudding

Serves 8

This decadent Maritime treat is a family favourite of my relatives in Lunenburg, Nova Scotia. Although, with proper planning, it can be enjoyed throughout the year, Maritime Cranberry Pudding tends to show up on tables during the fall and winter holiday time. Many households in eastern Canada include this special dessert as part of the Christmas dinner. The terrific thing about Maritime Cranberry Pudding is that it can be made in advance and frozen for use during those occasions. Provided your pudding is cooled completely first and is stored in a heavy, well-sealed container, it will keep up to six months in the freezer.

1 1/2 cups (375 mL) flour
1 tsp (5 mL) baking powder
1 cup (250 mL) fresh or frozen cranberries

1/2 cup (125 mL) molasses
2 tsp (10 mL) baking soda
1/2 cup (125 mL) boiling water

1/2 cup (125 mL) butter
1 cup (250 mL) granulated sugar
1 cup (250 mL) whipping cream

Preheat oven to 350°F (175°C). In large bowl, combine flour, baking powder and cranberries. Mix.

In medium bowl, combine molasses, baking soda and boiling water. Mix well. It will double in volume. Pour into flour mixture. Mix well. Pour batter into greased loaf pan. Bake in middle of oven with pan of water on bottom shelf of oven, for 50 to 60 minutes, until loaf is pulling away from sides of pan. Tip out of pan onto wire rack.

In medium saucepan over medium, combine butter, sugar and whipping cream. Bring to a soft boil, stirring occasionally. Serve warm sauce over warm pudding.

Whole-berry Cranberry Sauce

Makes 2 cups (500 mL)

Not only are cranberries steeped in history and tradition, but there is also a mystery surrounding the words "cranberry sauce" involving the Beatles and their famous "Paul is dead" hoax. During the recording of John Lennon's timeless classic "Strawberry Fields Forever," some banter between him and Ringo Starr was, mistakenly, heard as Lennon saying, "I buried Paul," when in fact Lennon said, "cranberry sauce." During an interview in 1980 just prior to his death, John Lennon admitted to occasionally throwing unrelated words and lyrics into his songs, such as "cranberry sauce." It just goes to show the historical importance of the celebrated cranberry.

1 cup (250 mL) water
1 cup (250 mL) granulated sugar
2 cups (500 mL) fresh or frozen cranberries

Combine water and sugar in medium saucepan over medium. Stir until dissolved. Add cranberries and cook for about 10 minutes, stirring occasionally, until they start to pop. Remove from heat and pour sauce into serving bowl. It will thicken as it cools. It will keep well in a sealed container in the refrigerator for up to 6 months.

Variation: Use 1 cup (250 mL) orange juice instead of water. The result will be a sauce with a fresh, citrusy zing.

Cran-apple Relish

Makes 3 cups (750 mL)

This traditional Cran-apple Relish recipe serves as the perfect fall or winter condiment and goes perfectly with North America's largest fowl, the turkey. The marriage of flavours between cranberry, apple and orange leaves a unique impression reminiscent of a cool, crisp autumn day. And don't let the whole orange rind component throw you off; after being ground in the food processor, you will never notice it, and the beautiful orange zest aroma will add a certain *je ne sais quoi* to an already terrific relish recipe.

2 tart apples
1 large, whole seedless orange
2 cups (500 mL) fresh cranberries
1 1/2 cups (375 mL) granulated sugar

Core and coarsely chop apples and place in food processor or grinder. Leave peel on orange and cut into sections. Add orange sections and cranberries to food processor. Process fruit into 1/2 inch (12 mm) pieces, and then transfer contents to medium bowl. Add sugar and mix well. Let sit for 1 hour before serving with turkey. It will keep well in a sealed container in the refrigerator for up to 3 months.

Cranberry Port Jelly

Makes 5 cups (1.25 L)

The majestic cranberry is steeped in history; it was a symbol of peace and friendship for several First Nations. One of the few truly native berries of North America, the cranberry was actually first named by the Pilgrims, who referred to it as "craneberry" because the fruit's blossom looked very much like the neck and head of a crane. Native peoples relied heavily on cranberries to help preserve meats and fish, and also used the berries in pemmican. Today we still often pair cranberries with meat, but we also like them sweet, and they are fantastic in this mildly sweet yet tangy jelly that goes great with roast meats but is also delicious on toast.

3 cups (750 mL) fresh (or frozen) cranberries
2 1/2 cups (625 mL) water
1 cup (250 mL) chopped unpeeled tart apple
1 x 1 inch (2.5 cm) piece ginger root, sliced
1 cinnamon stick
1 tsp (5 mL) whole black peppercorns

2 cups (500 mL) port wine
1 x 2 oz (57 g) box pectin crystals
2 Tbsp (30 mL) lemon juice

3 1/2 cups (875 mL) granulated sugar

Combine first 6 ingredients in Dutch oven. Bring to a boil, stirring occasionally. Reduce heat to medium-low. Simmer, covered, for about 30 minutes, crushing occasionally with a potato masher, until fruit is softened. Remove from heat. Carefully strain through colander lined with double layer of cheesecloth into large bowl. Let stand for 1 hour. Do not squeeze solids. Discard solids. Measure 1 1/2 cups (375 mL) liquid back into Dutch oven.

Add port wine, pectin crystals and lemon juice. Stir until pectin is dissolved. Bring to a boil, stirring constantly.

(continued on next page)

Add sugar. Bring to a hard boil, stirring constantly. Boil hard for 2 minutes, stirring constantly. Remove from heat. Carefully ladle hot mixture into five 1 cup (250 mL) hot sterile jars, leaving 1/4 inch (6 mm) headspace. Wipe rims clean. Place hot metal lids on jars and screw on metal bands fingertip tight. Process in boiling water bath for 15 minutes. Remove jars and allow to cool. Check that jars have sealed properly. Store sealed jars in a cool, dark place.

Hot Spiced Cranberry Drink

Serves 12

A perfect way to warm the cockles of your heart in our notoriously cold Canadian winters, this Hot Spiced Cranberry Drink is just what the doctor ordered! Whether you're spending the weekend at a scenic ski lodge or enjoying a New Year's celebration at the hunt camp, this aromatic cranberry drink will soon become a party favourite. It also makes the entire house smell wonderful when preparing it.

> **2 x 14 oz (398 mL) cans jellied cranberry sauce**
> **4 cups (1 L) pineapple juice**
> **3 cups (750 mL) water**
> **1/2 cup (125 mL) brown sugar**
> **1/2 tsp (2 mL) ground cinnamon**
> **1/2 tsp (2 mL) ground cloves**
> **1/4 tsp (1 mL) ground nutmeg**
> **1/4 tsp (1 mL) ground allspice**
> **1/8 tsp (0.5 mL) salt**
>
> **1 cup (250 mL) dark rum (optional)**

Measure first 9 ingredients into slow cooker. Stir well. Cover. Cook on Low for 4 hours.

Add rum. Stir. Serve hot.

Cranberry Lemon Sipper

Serves 8

Although it could be perceived as a "ladies drink," this Cranberry Lemon Sipper is a hit with many of the guys I know. The natural acidity and bite make it a real crowd pleaser. Start a batch of this cocktail in November so you'll have a ready supply to sip over the holidays. Serve as is, or mixed with ginger ale, tonic water or club soda for a festive spritzer.

4 cups (1 L) fresh (or frozen, thawed) cranberries
2 cups (500 mL) granulated sugar
1 cup (250 mL) water

2 cups (500 mL) sweet white wine
1 1/2 cups (375 mL) vodka
grated zest and juice of 1 lemon

In large saucepan, combine cranberries, sugar and water. Bring to a boil, stirring often. Reduce heat to medium. Boil gently, uncovered, for 5 to 10 minutes, stirring occasionally, until cranberries are softened. Remove from heat.

Add white wine, vodka and lemon zest and juice. Stir well. Cool. Pour into serile 10 cup (2.5 L) jar with tight-fitting lid. Let stand at room temperature for 3 weeks, shaking gently once every 2 days. Strain through sieve into large bowl. Do not press; gently lift cranberry mixture with spoon, allowing liquid to flow through sieve. Discard solids. Strain liquid again through double layer of cheesecloth into 8 cup (2 L) liquid measure. Pour into sterile jars or bottles with tight-fitting lids. Store at room temperature for up to 1 month. To serve, pour over crushed ice in small glass.

Festive Chicken Bake (p. 40)

Blackberry Pork Loin (p. 61)

Huckleberry Coffee Cake

Serves 8 to 10

Huckleberries grow in select parts of the Pacific Northwest and Rocky Mountain regions. A popular food source for wild animals of the region, huckleberries are also very popular with people; they are delicious freshly picked and can be used widely in cooking. Huckleberries are similar in taste to their cousin, the wild blueberry. The big difference is the seeds; fresh huckleberries have a crunchy texture. They are also a bit more tart than blueberries, with more intense flavouring. Certain recipes will call for huckleberries, and if you have access to them, it is worth giving them a shot.

1 cup (250 mL) flour
1 tsp (5 mL) baking powder
1/4 tsp (1 mL) salt

1/4 cup (60 mL) butter, softened
4 oz (125 mL) block cream cheese, softened
1 cup (250 mL) granulated sugar
1 egg
1 tsp (5 mL) vanilla extract
2 cups (500 mL) fresh or frozen huckleberries

2 Tbsp (30 mL) granulated sugar
1 tsp (5 mL) ground cinnamon

Preheat oven to 350°F (175°C). In small bowl, combine flour, baking powder and salt. Mix.

In medium bowl, beat butter and cream cheese together. Gradually add first amount of sugar; beat at medium speed until well blended. Beat in egg. Add flour mixture and mix well. Stir in vanilla. Fold in berries. Pour batter into greased 9 inch (23 cm) round baking pan.

Combine second amount of sugar and cinnamon, and sprinkle over batter. Bake for 1 hour, or until wooden pick inserted in centre comes out clean. Allow to cool before serving.

Huckleberry Pie

Serves 8

Because of its limited range in the wild, the huckleberry is somewhat less common than other Canadian berries; however, it has increased in popularity as people plant them in their own gardens. Huckleberry bushes require full sun and prefer cooler soil temperatures than their closely related cousin, the blueberry. A thick layer of protective mulch will keep their roots cool and conserve moisture. Be sure to give them room to grow; they will reach approximately 1 metre in height. They can also be cultivated in a large pot, as long as there is proper drainage and space for the roots. These bushes usually take at least one to two years before they will yield any berries, so patience is a virtue.

pastry for 2 crust, 9 inch (23 cm) pie

4 cups (1 L) fresh huckleberries
3/4 cup (175 mL) granulated sugar
1 Tbsp (15 mL) flour
1 tsp (5 mL) grated lemon zest
2 Tbsp (30 mL) lemon juice
2 Tbsp (30 mL) butter

2 Tbsp (30 mL) half-and-half cream
2 tsp (10 mL) granulated sugar

Preheat oven to 400°F (200°C). Line pie dish with pastry. Save remaining pastry for top crust.

Place huckleberries in pastry-lined dish. In small bowl, combine first amount of sugar and flour and mix well; spoon evenly over berries. Sprinkle lemon zest and lemon juice over top, then dot with butter. Cover with remaining pastry and press gently around edge to seal. Trim pastry edges and poke some holes randomly in top for steam vents.

Brush top crust with cream, and sprinkle with second amount of sugar. Bake for 15 minutes, then reduce oven temperature to 350°F (175°C). Continue baking for 20 to 25 minutes, until crust is golden brown.

Huckleberry

Huckleberry Relish

Makes 3 cups (750 mL)

Few people realize that the powerful grizzly bear actually feeds extensively on berries throughout the growing season. Along with blueberries, salmonberries, and soapberries (also known as buffalo berries), the omnivorous grizzly forages extensively on huckleberries, depending on crop productivity and seasonal changes that year. Some biologists estimate that plants and berries compose 80 percent of the grizzly's diet in many parts of the Pacific Northwest, contrary to the popular belief that these bruins eat mostly meat.

1 cup (250 mL) water
2 cups (500 mL) fresh or frozen huckleberries
1/2 cup (125 mL) granulated sugar

1 onion, thinly sliced
1/2 cup (125 mL) apple cider vinegar
4 garlic cloves, minced

Combine water, huckleberries and sugar in heavy-bottomed saucepan and bring to a boil over medium. Cook, stirring, until berries soften, about 5 minutes. Remove from heat.

Add onion, cider vinegar and garlic. Mix well. Cover and allow to cool. Ladle into jars, screw on lids and refrigerate. It will keep well in the refrigerator for up to 3 months.

Blackberry Breakfast Bars

Makes 18 bars

My wife and I have been picking blackberries together for nearly 20 years. We first began searching for the beloved blackberry in the woods behind her family cottage in the Laurentian Mountains. I can recall the excitement of stumbling upon a patch at perfect ripeness; it was like winning the lottery. Now we annually search the hills and dales around our campground up the Ottawa Valley, where blackberries hide amongst the other wild edibles. There is nothing like picking berries to bring a family closer together.

2 cups (500 mL) fresh or frozen blackberries
2 Tbsp (30 mL) granulated sugar
2 Tbsp (30 mL) water
1 Tbsp (15 mL) lemon juice
1/2 tsp (2 mL) ground cinnamon

1 cup (250 mL) flour
1 cup (250 mL) rolled oats
2/3 cup (150 mL) brown sugar
1/4 tsp (1 mL) ground cinnamon
1/8 tsp (0.5 mL) baking soda
1/2 cup (125 mL) butter, melted

To make filling, combine blackberries, granulated sugar, water, lemon juice and first amount of cinnamon in saucepan over medium-high. Bring to a boil, then reduce heat and allow to simmer, uncovered, stirring frequently, for about 8 minutes, until slightly thickened. Remove from heat.

Preheat oven to 350°F (175°C). In large bowl, combine flour, rolled oats, brown sugar, second amount of cinnamon and baking soda, and mix well. Stir in melted butter until well blended. Set aside 1 cup (250 mL) of oat mixture for topping. Press remaining oat mixture into ungreased 9 x 9 inch (23 x 23 cm) baking pan. Bake for 20 to 25 minutes, until crust is browned. Carefully spread filling over baked crust. Sprinkle with reserved oat mixture. Lightly press oat mixture into filling. Bake for another 20 to 25 minutes, until topping browns. Allow to cool, then cut into bars.

Easy-peasy Oatmeal Raspberry Swirl

Serves 1

Raspberries are grown and picked extensively in North America, but the raspberries we have here in Canada pale in comparison to the raspberries in Poland, the world's largest raspberry-producing country. In 2013, Poland harvested a staggering 127,000 tons of raspberries, which was a record over the previous year. This is compared to 511 tons produced here in Canada. According to berry experts, the high-quality Polish raspberry has to do with the ideal soil in that country combined with perfect climate and advanced technological practices. You could say that the Polish are the raspberry experts of the world.

> 1/3 cup (75 mL) instant rolled oats
> 3/4 cup (175 mL) milk
>
> 2 Tbsp (30 mL) raspberry jam (see Note)
> 2 Tbsp (30 mL) whipped topping
> 5 fresh raspberries

Combine rolled oats and milk in microwave-safe bowl and cook in microwave for 1 1/2 to 2 minutes, stirring after 1 minute.

Swirl jam through warm oatmeal. Serve in cereal bowl with a dollop of whipped topping. Top with fresh raspberries.

Note: If you don't want to use store-bought jam, you can make your own; Raspberry Honey Jam, page 79, will work well in this recipe.

Spicy Raspberry Spread

Serves 6 to 8 as an appetizer

If you're ever visiting Newfoundland in summer, do yourself a favour and check out the lovely East Coast Trail. Everything from blueberries and cranberries to blackberries and raspberries can be found along this scenic walking path. Although "the Rock" is not historically known as a hub for wild berries, there are certain pockets worth investigating. I would recommend Stiles Cove Path—a 15 kilometre hiking trail that links Flatrock to Pouch Cove—as an excellent spot to forage for fresh raspberries. You will find copious raspberries in any place where the forest has been previously cleared, as is the case along certain sections of Stiles Cove Path.

> 1 cup (250 mL) raspberry preserves (see Note)
> 1 cup (250 mL) picante sauce
> 1 x 8 oz (250 g) block cream cheese

In medium bowl, mix together raspberry preserves and picante sauce until smooth. Place cream cheese block on serving dish. Cover with raspberry mixture. Serve with your choice of crackers.

Note: If you don't want to use store-bought preserves, you can make your own; Raspberry Preserves, page 78, will work well in this recipe.

Melon Raspberry Soup

Serves 6 as an appetizer

Be warned that this is not the soup your grandmother used to make. Don't fret, though; Melon Rasperry Soup may just be the best concoction to hit your bowl in a long time. I must admit that a gazpacho-style soup was not on my bucket list of things to eat, but I was pleasantly surprised when it turned out so well. The first time I tried it, I pictured myself dining in some small restaurant in the South of France. I even considered sporting a beret. Try it; the clean, cool taste and fragrance combine for a sensory experience you will not soon forget.

> 5 cups (1.25 L) chopped ripe cantaloupe
> 1 cup (250 mL) white grape juice
> 2 Tbsp (30 mL) lemon juice
>
> 1 1/2 cups (375 mL) fresh (or frozen, thawed) raspberries
> 1/3 cup (75 mL) white grape juice

Process cantaloupe, first amount of grape juice and lemon juice in blender or food processor until smooth. Transfer to medium bowl. Chill, covered, for at least 2 hours, until cold.

Meanwhile, process raspberries and second amount of grape juice in blender or food processor until smooth. Strain through fine-mesh sieve into small bowl. Discard seeds. Chill, covered, for at least 2 hours, until cold. Drizzle raspberry sauce over individual servings of melon soup.

Venison Steak with Blackberry Compote

Serves 4

The majority of all deer meat found in wild game specialty shops, or taken by hunters and served to family and friends, is that of the male or buck. But as part of sound wildlife management, a certain number of antlerless deer must also be harvested each year to achieve a population balance. The meat of a young deer or fawn is comparable to veal in beef, and the meat of a female deer or doe is similar to the buck, although, depending on the age of the deer, it has a tendency to be more tender. No matter the age or whether male or female, venison is delicious and perhaps the healthiest protein you'll ever consume, especially when served with blackberries.

2 Tbsp (30 mL) minced shallots
1 tsp (5 mL) minced garlic
3 Tbsp (45 mL) blackberry jam (see Note)
1 cup (250 mL) dry red wine

1 cup (250 mL) prepared beef broth
1 Tbsp (15 mL) butter
1/4 tsp (1 mL) salt
1/4 tsp (1 mL) pepper

2 Tbsp (30 mL) vegetable oil
4 venison steaks
12 to 16 fresh blackberries

In large saucepan over medium-high, heat shallots, garlic, blackberry jam and red wine. Simmer for about 15 minutes, until reduced to about 1/2 cup (125 mL) liquid. Strain liquid through fine-mesh sieve and set aside.

In separate saucepan over medium-high, simmer beef broth for 15 to 20 minutes, until reduced by half. Whisk wine sauce into reduced broth, and stir in butter. Season with salt and pepper.

Heat oil in skillet over medium-high. Cook venison steaks for 3 to 5 minutes per side, until they are beginning to firm and are hot and slightly pink in centre. Serve steaks with sauce and a few fresh blackberries.

Note: If you don't want to use store-bought jam, you can make your own; Blackberry Jam, page 80, will work well in this recipe.

Blackberry Pork Loin

Serves 6

Blackberries are not only fun to forage and delicious to eat, but they will also keep your mouth in good working order. The blackberry, along with its bark and leaves, can be eaten and used as an astringent. The tannins in blackberries help prevent inflammation of the gums and play a part in good overall hygiene. Aboriginals discovered this useful property long ago. Some groups concocted a mouthwash using blackberry, which they would gargle with to help with oral hygiene. Eating blackberries may never become a replacement for brushing your teeth, but they certainly might help keep the dentist away.

2 1/2 lbs (1.1 kg) boneless pork loin roast
2 Tbsp (30 mL) olive oil
salt and pepper

2 cups (500 mL) fresh or frozen blackberries
1/4 cup (60 mL) granulated sugar
1/4 cup (60 mL) water
2 Tbsp (30 mL) apple brandy

1/2 cup (125 mL) apple cider
2 Tbsp (30 mL) grainy mustard

Preheat oven to 450°F (230°C). Place pork loin roast on rack in roasting pan and coat with oil, then season to taste with salt and pepper. Place in oven and cook for 10 minutes. Reduce heat to 350°F (175°C) and cook for 20 minutes more.

In medium saucepan, combine blackberries, sugar and water. Bring to a boil. Deglaze pan with apple brandy for about 5 minutes, until alcohol is cooked off.

Add apple cider and mustard. Stir. Simmer sauce for about 10 minutes, until reduced by three-quarters. Brush glaze on pork loin and continue to roast for another 10 to 15 minutes, until internal temperature reaches 150°F (65°C). Let pork loin rest, covered, for 15 minutes. Slice pork and drizzle sauce over slices before serving.

Pictured on page 52.

Poached Chicken with Blackberry Cabernet Sauce

Serves 6

Consuming 1 cup (250 mL) of these terrific berries serves as half of the daily recommendation of vitamin C. Our immune systems use vitamin C to help ward off illness, and regular intake of vitamin C will also lower the risk of heart disease. That 1 cup of blackberries also contains over 30 percent of the daily recommended amount of fibre, which is essential for healthy digestion.

2 cups (500 mL) water
1 cup (250 mL) dry white wine
1 bay leaf
1 small onion, chopped
1 large carrot, chopped
leaves from 1 celery rib, chopped
1/2 tsp (2 mL) salt
1/4 tsp (1 mL) pepper
2 lbs (900 g) bone-in chicken breast halves

2 cups (500 mL) fresh (or frozen, thawed) blackberries

3 Tbsp (45 mL) extra-virgin olive oil
1 small onion, chopped
3 Tbsp (45 mL) granulated sugar
1 cup (250 mL) Cabernet Sauvignon wine

In large pot over medium, combine water, white wine, bay leaf, first onion, carrot, celery leaves, salt and pepper. Bring to a boil and cook for 5 minutes. Add chicken. Bring back to a boil, cover, reduce heat to low, and simmer for 15 minutes. Turn off heat and let pot stand, covered, until cool.

For sauce, process blackberries in food processor or blender until smooth. Strain enough blackberry purée to measure 1 cup (250 mL) seedless pulp.

In large frying pan over medium-high, heat olive oil. Add second onion and cook for about 5 minutes, until transparent. Stir in blackberry pulp, sugar and Cabernet Sauvignon. Simmer for 10 to 12 minutes, until reduced by half. Remove from heat. Remove chicken pieces from pot. Skin and debone breasts and slice thinly lengthwise. Arrange sliced breast on a platter, top with sauce and serve.

Raspberry Pistachio Chicken

Serves 2

As you put together this scrumptious Raspberry Pistachio Chicken, consider that this sought-after nut requires a very long growing season in a temperate climate. The pistachio is grown almost exclusively in the hot climates of Iran and the United States, so recent political upheaval in the Middle East has had a direct effect on pistachio supply. Growers in California stand to benefit greatly from a U.S. embargo on the import of pistachios from Iran. But without getting too political, we Canadians are okay with having to purchase American-grown pistachios.

> 1/2 cup (125 mL) fresh (or frozen, thawed) raspberries
> 1 tsp (5 mL) Dijon mustard
> juice of 1 lemon
>
> 1/4 cup (60 mL) dry bread crumbs
> 1 Tbsp (15 mL) chopped pistachios
> 1 Tbsp (15 mL) parsley flakes
> 1/8 tsp (0.5 mL) salt
> 1/8 tsp (0.5 mL) pepper
> 2 boneless, skinless chicken breast halves
>
> 1 Tbsp (15 mL) olive oil

Combine raspberries, mustard and lemon juice in blender or food processor and process until smooth. Transfer to shallow dish.

In another dish, combine bread crumbs, pistachios, parsley, salt and pepper. Dip chicken into raspberry sauce, then into pistachio mixture to coat.

Heat olive oil in skillet over medium. Add chicken and cook for about 8 minutes per side, until done.

Delectable Duck Breasts with Raspberry Sauce

Serves 4

As with some other waterfowl, duck is a somewhat fatty meat with a thick layer of skin that becomes quite crispy when grilled or broiled. Compared to upland game birds, duck breasts contain a lot of natural oils, which will cook off in the first few minutes. If you prefer not to cook your duck in its own juices, you can drain off the grease in the early stages of the cooking process or remove the skin altogether. Another technique some chefs use is to make small slices into the breast meat approximately 1 inch (2.5 cm) apart. This allows grease to drain out, thereby improving the overall taste of your bird.

4 duck breast halves

1/2 cup (125 mL) dry red wine
3 Tbsp (45 mL) crème de cassis liqueur
1 tsp (5 mL) cornstarch
1/2 cup (125 mL) fresh (or frozen, thawed) raspberries

1 Tbsp (15 mL) brown sugar
2 tsp (10 mL) ground cinnamon
2 tsp (10 mL) salt

Use fork to pierce duck breasts through skin and fat but not all the way through to meat. Place breasts in large, heavy skillet on medium-high and cook, skin side down, for about 10 minutes, until skin browns and fat runs out. Remove pan from heat and pour off fat. Return to heat and fry breasts, skin side up, for another 10 minutes. Remove breasts from pan and allow to rest on baking sheet. Pour most of fat out of skillet, leaving only a coating.

In small bowl, combine red wine, liqueur and cornstarch and mix well. Pour mixture into skillet and simmer for 3 minutes, stirring constantly, until sauce is thickened. Add raspberries and simmer for 1 minute, until heated through.

Preheat broiler. In another small bowl, combine brown sugar, cinnamon and salt. Sprinkle over skin of duck breasts. Broil for about 1 minute, until sugar begins to caramelize.

Slice duck breasts thinly, pour sauce over top and serve.

Pictured on page 69.

Berry-glazed Salmon

Serves 2

Salmon have become a common resident of the Great Lakes in recent years.
Anglers come from far and wide to fish Lake Ontario in hopes of hooking into
a good-sized Chinook or Coho. I can recall a charter fishing trip I did out of
Bath, Ontario, near Kingston. My friends and I caught many lake trout but only
a few salmon. It wasn't difficult to tell when you had a salmon on the line. If
your rod bent over and bounced slightly when down-rigger went off, it was
a lake trout, but if the rod doubled over and the drag started "screaming," you
knew you had a salmon. Salmon are one of the hardest-fighting fish in the
world, which I learned first hand on that trip.

3/4 cup (175 mL) fresh blackberries
3/4 cup (175 mL) fresh raspberries
2 Tbsp (30 mL) dry red wine
2 Tbsp (30 mL) brown sugar

2 x 7 oz (200 g) salmon fillets or steaks
salt and pepper

Set aside 1/4 cup (60 mL) each of blackberries and raspberries. Combine
remaining berries and wine in small saucepan and heat to a light simmer,
stirring and crushing berries until juicy, about 5 minutes. Strain berries;
discard pulp and seeds. Juice should measure 1/2 to 2/3 cup (125 to 150 mL).
Return juice to saucepan. Add brown sugar and continue to simmer until
mixture is syrupy and reduced to about 1/3 cup (75 mL).

Preheat broiler and place rack 6 to 8 inches (15 to 20 cm) from heat source.
Line baking sheet with foil and coat with cooking spray. Season salmon with
salt and pepper and place on baking sheet. Broil for 10 minutes, flipping
at the halfway point, until nearly cooked through. Brush tops of fillets
with 1 Tbsp (15 mL) berry syrup and broil for another 2 minutes. Stir
remaining berries into berry syrup and reheat briefly. Serve salmon with
warm berry sauce.

Raspberry Cobbler

Serves 6 to 8

Because raspberry seeds are a nuisance to chew and tend to get caught in your teeth, some people go to great lengths to use the raspberry without all the seeds. There are several techniques for eliminating the seeds in raspberries if you are not a fan. One method is to push the pulp through a fine-mesh sieve for the juice without the seeds. Another method is to use a food mill in conjunction with a series of mesh screens. The screens will capture the tiny seeds from the raspberries. Finding the proper mesh size in these special strainers can be difficult; occasionally, even the finest mesh will not catch all the seeds.

1 cup (250 mL) flour
1/2 cup (125 mL) granulated sugar
1 tsp (5 mL) baking powder
1/2 tsp (2 mL) salt
6 Tbsp (90 mL) cold butter
1/4 cup (60 mL) boiling water

2 Tbsp (30 mL) cornstarch
1/4 cup (60 mL) cold water
1 cup (250 mL) granulated sugar
1 Tbsp (15 mL) lemon juice
4 cups (1 L) fresh raspberries

Preheat oven to 400°F (200°C). In large bowl, mix flour, first amount of sugar, baking powder and salt. Cut in butter until mixture resembles coarse crumbs. Stir in boiling water just until dough is evenly moist.

In separate bowl, dissolve cornstarch in cold water. Mix in second amount of sugar, lemon juice and raspberries. Transfer mixture to skillet and bring to a boil, stirring frequently. Drop dough into skillet by the spoonful. Place skillet on baking sheet to avoid it bubbling over in oven. Bake for 25 minutes, or until dough is golden brown.

Pictured on page 70.

Quick Blackberry Cobbler

Serves 6 to 8

Alberta is dedicated to promoting its local food crops. The Alberta Farm Fresh
Producers Association (AFFPA) was established to support farm direct markets
for vegetable crops, berry and fruit crops, and more. The AFFPA has voluntary
membership, is non-profit and represents direct market growers throughout the
province. The main goal of the AFFPA is to support and grow a profitable and
sustainable farm fresh products market. As a member, you can get involved
in an association chapter, sign up for their newsletter and help promote
sustainable farm fresh products throughout Alberta.

1/2 cup (125 mL) butter, melted

1 cup (250 mL) granulated sugar
1 cup (250 mL) self-rising flour
3/4 cup (175 mL) milk
2 cups (500 mL) fresh or frozen blackberries

Preheat oven to 350°F (175°C). Pour melted butter into 8 x 8 inch
(20 x 20 cm) baking pan and spread to cover bottom.

In medium bowl, mix sugar, self-rising flour and milk together until
moistened. Pour mixture evenly over butter; do not stir. Spread blackberries
evenly over batter. Bake for about 45 minutes, until top is browned and
cobbler is bubbling.

Blackberry Pie

Serves 8

For anyone who has not visited Canada's beautiful province of Nova Scotia, I recommend it, especially for outdoor enthusiasts and foraging freaks. The province's gorgeous Annapolis Valley is the perfect place to start. For the avid berry picker who enjoys camping, check out Blomidon Provincial Park or Valleyview Provincial Park, or some of the beautiful nature trails such as Balancing Rock, Cape Blomidon and the Kentville Agricultural Research Station trail. Many parts of the Annapolis Valley are berry-picking paradise during the summer months. And what better way to use those berries than in a delicious pie.

pastry for 2 crust, 9 inch (23 cm) pie

4 cups (1 L) fresh blackberries
3 Tbsp (45 mL) flour
1 cup (250 mL) granulated sugar
1 Tbsp (15 mL) lemon juice
1 Tbsp (15 mL) butter

Preheat oven to 400°F (200°C). Line pie dish with pastry. Save remaining pastry for top crust.

Combine blackberries, flour, sugar and lemon juice. Mix well. Spoon filling into pie shell and dot with butter. Cover with remaining pastry and press gently around edge to seal. Trim pastry edges and poke some holes randomly in top for steam vents. Bake for 15 minutes, then reduce oven temperature to 350°F (175°C). Continue baking for 35 to 40 minutes, until crust is golden brown.

Delectable Duck Breasts with Raspberry Sauce (p. 64),
Cherry Cranberry Martini (p. 137)

Raspberry Cobbler (p. 66)

Chocolate Raspberry Torte

Serves 8

Torte is a variety of cake originating in Europe. It is extremely popular in France and Germany. Tortes are typically multi-layered, filled with cream, jam or fruit and often have a glaze as opposed to icing. Occasionally they will be made without the layering. Tortes are generally baked in a springform pan. To the average person, a torte looks very much like any other cake until you examine the subtle differences more closely.

12 x 1 oz (28 g) semi-sweet chocolate baking squares, *divided*
3/4 cup (175 mL) butter

1 x 3 oz (85 g) box raspberry jelly powder
1/2 cup (125 mL) granulated sugar
3 eggs
1/3 cup (75 mL) flour

2 cups (500 mL) frozen whipped topping, thawed
1/2 cup (125 mL) fresh raspberries
1 Tbsp (15 mL) icing sugar

Preheat oven to 350°F (175°C). Coarsely chop 4 chocolate squares and place them and butter in large microwave-safe bowl. Heat on medium for 1 minute. Stir. Return to microwave for 1 minute, or until chocolate is almost melted. Stir until chocolate is completely melted.

Add jelly powder and granulated sugar; mix well. Beat in eggs. Add flour; mix well. Pour into greased and parchment paper-lined 9 inch (23 cm) round baking pan. Bake for 40 minutes, or until wooden pick inserted in centre comes out with fudgy crumbs. Allow cake to cool in pan for 5 minutes. Run small knife around edge of cake to loosen. Turn upside down and place on wire rack; remove parchment paper. Let cake cool completely.

In another microwave-safe bowl, melt another 4 squares of chocolate as before. Spread melted chocolate onto parchment paper-lined baking sheet and refrigerate for 30 minutes. Melt remaining 4 squares of chocolate in large microwave-safe bowl as before. Whisk in whipped topping until well blended. Transfer cake to plate; frost with whipped topping mixture. Take chocolate out of refrigerator and break into irregular bite-sized pieces and arrange on cake. Top with raspberries and icing sugar.

Raspberry White Chocolate Waffle Pudding

Serves 4

Wild raspberries are distinctive and readily available in many parts of Canada. My wife and I have picked more raspberries in our day than I could ever count. At her family cottage near Mont Tremblant in the Laurentians, wild red raspberries grow in huge quantities. Raspberry bushes are often the first vegetation to reappear in heavily logged areas. The young growth of raspberries serves as some consolation for the loss of our great forests, especially during years of bumper crops. If you have the patience and some good protective clothing, you can easily forage a year's supply of berries.

8 fresh or frozen waffles
1 cup (250 mL) fresh or frozen raspberries
4 x 1 oz (28 g) white chocolate baking squares, chopped

2 Tbsp (30 mL) caster (or granulated) sugar
1 Tbsp (15 mL) flour
2 eggs
1/2 tsp (2 mL) grated lemon zest
1/2 tsp (2 mL) vanilla extract
1 cup (250 mL) half-and-half cream
1 Tbsp (15 mL) icing sugar

Preheat oven to 350°F (175°C). Cut waffles into small cubes and place half of cubes in greased 1 1/2 quart (1.5 L) casserole. Place half of raspberries on top of waffles, then half of chocolate on top of raspberries. Repeat 3 layers.

In small bowl, whisk together caster sugar, flour, eggs, lemon zest, vanilla and cream. Pour over waffles and let stand for 10 minutes. Bake for about 35 minutes, until golden. Dust with icing sugar and serve.

Lemon Berry Mousse

Serves 6

In the wild, the black raspberry is often mistaken for the blackberry. Both red and black raspberries grow in the wild, and there are several ways to tell the difference between a black raspberry and a blackberry bush. Raspberry canes are usually twice as tall as blackberry canes. Black raspberry canes also bend back toward the ground. The stems have a pale, almost bluish colour that rubs off if you touch them. Blackberry canes tend to be lower to the ground. The easiest way to tell the difference is by picking the berries. Raspberries are hollow when picked; blackberries come off the bush with their core intact.

1 x 11 oz (300 mL) can sweetened condensed milk
1 Tbsp (15 mL) grated lemon zest
1/3 cup (75 mL) lemon juice
1 cup (250 mL) plain yogurt
2 cups (500 mL) frozen whipped topping, thawed, *divided*

2/3 cup (150 mL) fresh blackberries
2/3 cup (150 mL) fresh raspberries

Combine condensed milk, lemon zest and lemon juice and mix until smooth. Stir in yogurt. Fold in 1 1/2 cups (375 mL) whipped topping. Divide mixture among 6 dessert bowls. Chill for 3 to 4 hours.

Just before serving, top with berries and remaining 1/2 cup (125 mL) whipped topping.

Chocolate Raspberry Truffles

Makes 32 to 48 chocolates

The first time my wife made chocolate raspberry truffles for me, I fell in love with her all over again. It was the most decadent dessert I had ever eaten, and we still treat ourselves on occasion by whipping up a beautiful batch. They are somewhat on the labour intensive side; however, if you want to impress some dinner guests, give these truffles a shot. Trust me—your guests will come back for more. Just be warned, this recipe seems to have the cupid effect on most people.

1 1/2 cups (375 mL) fresh (or frozen, thawed) raspberries
1/4 cup (60 mL) icing sugar

1 lb (454 g) semi-sweet chocolate baking squares (16 squares), chopped
3/4 cup (175 mL) cream
2 Tbsp (30 mL) white corn syrup

1 lb (454 g) chocolate melting wafers

1/4 cup (60 mL) red candy melting wafers (optional)

Place raspberries in blender or food processor and process until smooth. Pour raspberry purée into small saucepan through fine-mesh sieve or cheesecloth to remove seeds. Add icing sugar to raspberry purée and heat over medium, stirring frequently, until thick and syrupy and reduced by about half. Remove from heat and set aside.

To make ganache, place semi-sweet chocolate in large bowl. Heat cream in small saucepan over medium until bubbles start to form around edges, but do not allow it to come to a full boil. Pour hot cream over chocolate and let sit for 1 to 2 minutes to soften chocolate. Gently whisk chocolate and cream together, until chocolate is melted and mixture is smooth. Add corn syrup and raspberry reduction to chocolate mixture, and whisk together. Cover surface of ganache with cling wrap, and refrigerate for about 3 hours, until thick enough to scoop. Using candy scoop or teaspoon, scoop ganache into small balls and place on foil- or waxed paper-lined baking sheet. Once all of ganache is scooped, freeze balls until firm, about 2 hours.

(continued on next page)

Place chocolate melting wafers in large microwave-safe bowl. Heat in 1-minute increments, stirring after every minute to prevent overheating, until chocolate is melted and completely smooth. Allow to cool slightly so that it does not melt ganache balls. Using a fork, dip 1 ganache ball into melted chocolate. Allow excess to drip back into bowl. Place dipped truffle back onto baking sheet and repeat with remaining ganache balls and chocolate. Place truffles in refrigerator for about 30 minutes to set chocolate coating.

If you wish to decorate truffles, melt red candy wafers and drizzle over truffles. Truffles can be stored in refrigerator for up to 2 weeks.

Raspberry Buttercream Icing
Makes 2 cups (500 mL)

Berries have the effect of adding natural colour as well as flavour. Raspberries and blueberries make an excellent food dye. Colour changes brought about by berries will be subtle, providing a nice hue or tint to different icings as well as in other recipe choices. This icing turns out a beautiful shade of pink. Why use food colouring when you can provide colour the natural way?

1/2 cup (125 mL) butter, softened
1/2 cup (125 mL) fresh (or frozen, thawed) raspberries
1 tsp (5 mL) vanilla extract
1/8 tsp (0.5 mL) salt

2 cups (500 mL) icing sugar

In medium bowl, combine butter, raspberries, vanilla and salt. Using electric mixer on medium speed, beat mixture until creamy.

Reduce speed to low. Gradually add icing sugar, beating until blended and smooth after each addition. Use icing to frost chocolate or vanilla cake or cupcakes.

Raspberry Southwest Chipotle Sauce

Makes 2 cups (500 mL)

The sweet acidity of raspberries brings a perfect balance to the zip and heat of jalapeño peppers. The resulting sauce is the perfect combination of sweet and savoury and something you can use to complement a variety of different meats. A flood of chipotle-style sauces has hit the market in recent years, but this homemade one happens to be my favourite. It was actually my eldest daughter, Emily, who first introduced our family to the joys of a Southwest-style chipotle, and we worked at finding a recipe to make our own sauce that would deliver some light summertime tones. I think we've succeeded with this one, but you be the judge.

2 Tbsp (30 mL) olive oil
2 large jalapeño peppers, seeded and diced

2 garlic cloves, minced
1 Tbsp (15 mL) chili powder
1 1/2 cups (375 mL) fresh (or frozen, thawed) raspberries

1/2 cup (125 mL) apple cider vinegar
1/2 cup (125 mL) granulated sugar
1/4 cup (60 mL) brown sugar
1/2 tsp (2 mL) salt

Heat olive oil in skillet over medium. Stir in jalapeño and cook for about 5 minutes, until tender.

Add garlic and chili powder and bring to a simmer. Stir in raspberries and cook for about 3 minutes, until soft.

Stir in vinegar, both sugars and salt. Simmer for about 15 minutes, until thickened and reduced by half. Serve over salmon, chicken or pork. It will keep well in a sealed container in the refrigerator for up to 3 months.

Raspberry Vinaigrette

Makes 2 cups (500 mL)

Many of our most popular Canadian berries made *Canadian Living* magazine's feature on the 25 most nutritious fruits and berries. Raspberries, strawberries, blueberries, blackberries, cranberries and even cherries ranked very high on the list. The article looked at nutritional value and antioxidant value along with other elements such as disease-fighting properties, all present at high levels in berries. The article describes fruit and berries so eloquently as "a complex combination of fibre, minerals, antioxidants and phytochemicals—as well as the vitamins—that work in combination to provide protective benefits."

1 cup (250 mL) fresh (or frozen, thawed) raspberries
1/2 cup (125 mL) apple cider vinegar
1/2 cup (125 mL) balsamic vinegar
2 tsp (10 mL) granulated sugar (see Note)
1 1/2 Tbsp (22 mL) Dijon mustard
1/2 cup (125 mL) extra-virgin olive oil

Process raspberries, both vinegars, sugar and Dijon mustard in blender until smooth. Add olive oil slowly until combined. Toss with your favourite salad ingredients. It will keep well in a sealed container in the refrigerator for up to 6 months.

Note: Depending on the tartness of your raspberries, you may need to adjust the amount of sugar you add. If you find the vinaigrette is too tart, add an extra 1/2 tsp (2 mL) or so of sugar.

Raspberry Preserves

Makes 4 cups (1 L)

This recipe is a quick and easy way to preserve raspberries, allowing you to store them to enjoy at a later date. Should you have a large amount of berries, simply multiply the recipe for similar results. You may also substitute other fruits during the busy berry season when several types of berries are ripening at the same time. Preserves are typically bottled these days, as with this recipe, but years ago, they were canned and stored in other sealed containers for use throughout the cooler or leaner months.

4 cups (1 L) fresh raspberries

Syrup Proportion:
1 cup (250 mL) granulated sugar
2 cups (500 mL) water

Pack raspberries into two 2 cup (500 mL) jars, pressing down firmly without crushing fruit.

Fill one jar of fruit with water. Drain water into liquid measuring cup. Multiply the amount by 2, and you will know how much syrup to make. Measure sugar and water into saucepan. Heat on medium-high, stirring occasionally, until it boils. Pour over raspberries to within 1/2 inch (12 mm) of top of each jar. Wipe rims clean. Place hot metal lids on jars and screw on metal bands fingertip tight. Process in boiling water bath for 15 minutes. Remove jars and allow to cool. Check that jars have sealed properly. Store sealed jars in a cool, dark place.

 tip   **USING A BOILING WATER CANNER**

Set the canner on stove element heat on medium-high; fill half full of water. Fill each jar and place it in rack, replacing canner cover each time. After adding the last jar, add boiling water to reach 1 inch over jar tops. Cover and heat. Begin timing when water is boiling. Keep water boiling gently, adding more boiling water if level drops. If water stops boiling when more is added, stop timing, turn up heat, and wait for a full boil before resuming timing. At end of the processing time, turn off heat and remove jars to cool.

Raspberry Honey Jam

Makes 4 cups (1 L)

You will notice the use of honey in this recipe instead of granulated sugar.
As such a wonderful source of natural sugar, this product of our busy bees gives
a great natural touch to homemade jam. Honey is one of the many wonders
of nature, and for survivalists, honey is a staple of life. It can last forever
without the need of refrigeration. Even when your honey solidifies, it is still
usable. Simply heat it up and it will return to its state of free-flowing goodness.
Ah, honey—one of nature's most perfect foods.

4 cups (1 L) fresh raspberries, *divided*

1 x 2 oz (57 g) box pectin crystals
1 cup (250 mL) honey (adjust for more or less sweetness)
1 cup (250 mL) unsweetened apple juice

In medium bowl, crush 2 cups (500 mL) raspberries. In separate bowl, crush
remaining 2 cups (500 mL) raspberries, and press through fine-mesh sieve.

Place all raspberries in large saucepan and whisk in pectin until dissolved.
Bring to a boil over medium, stirring constantly. Stir in honey and bring
to a boil again. Stir in apple juice and bring to a boil. Boil for 3 minutes,
stirring occasionally to keep from burning. Remove from heat. Skim off
any foam. Carefully ladle hot mixture into hot sterile jars, leaving 1/4 inch
(6 mm) headspace. Wipe rims clean. Place hot metal lids on jars and screw
on metal bands fingertip tight. Process in boiling water bath for 15 minutes.
Remove jars and allow to cool. Check that jars have sealed properly. Store
sealed jars in a cool, dark place.

Blackberry Jam

Makes 2 cups (500 mL)

Although most of us are aware of the important role that wild berries play in the Canadian wildlife food chain, few of us realize exactly to what extent berries play in the diets of large omnivores such as the black bear. Wild berries are so important to bears that they dictate the bears' movement throughout the year. In years when berry crops are strong, black bears usually remain deep in the forest with plenty to eat, but in years when berry crops are down, governments have documented marked increases in nuisance bear activity as the animals search for alternate food sources.

4 cups (1 L) fresh blackberries
1 cup (250 mL) granulated sugar
2 Tbsp (30 mL) cornstarch
1/4 tsp (1 mL) ground cinnamon

Place blackberries in large saucepan and mash. Stir in sugar until juices form. Place about 1 Tbsp (15 mL) blackberry juice in small bowl and stir in cornstarch. Add cornstarch mixture to blackberry mixture. Bring to a boil over medium-high, stirring often, until jam is thickened, about 15 minutes. Stir in cinnamon. Remove from heat and allow to cool. Transfer jam to two 1 cup (250 mL) jars. Cover and refrigerate.

Traditional Blackberry Syrup

Makes 6 cups (1.5 L)

Blackberries are some of the most delectable of all wild berries and one of my personal favourites. Recently, however, scientists have been baffled by a fungal disease called rust, found in both blackberries and raspberries. Rust tends to set in during wet conditions and can adversely affect plant growth. While foraging for wild blackberries on crown or public land, if you notice any orange growth on the plant stock, cut the entire plant away down to the roots. Rust can spread through the root system from plant to plant, and once a plant is infected, it is infected for life.

10 cups (2.5 L) fresh blackberries

5 cups (1.25 L) granulated sugar
1/3 x 2 oz (57 g) box pectin crystals

Place blackberries in large saucepan and mash. Heat just enough to warm them a bit so juice runs better. Strain berries through jelly cloth or cheesecloth. Do not squeeze bag. Measure 4 cups (1 L) juice back into saucepan.

Add sugar. Mix well and bring to a boil over medium-high. Allow to boil rapidly for 2 minutes. Skim off any foam. Allow to cool, and then add pectin and mix. Pour into three 2 cup (500 mL) hot sterile jars, leaving 1/4 inch (6 mm) headspace. Wipe rims clean. Place hot metal lids on jars and screw on metal bands fingertip tight. Store jars in refrigerator.

Blackberry Lime Soda

Serves 4

When I first heard about blackberry soda, I thought it was a joke. "You mean you can actually make your own soda flavours?" I said to myself. It wasn't until I gave it a shot that I discovered how truly great Blackberry Lime Soda really was. I usually put it together with club soda, but lemon lime pop creates a sweeter version. Either way, blackberry is an added twist to something you already enjoy. Garnish glasses with fresh berries or lime wedges if desired.

> 1 1/3 cups (325 mL) fresh blackberries
> 1/2 cup (125 mL) lime juice
> 3 Tbsp (45 mL) granulated sugar
>
> ice cubes, for serving
> 4 cups (1 L) club soda (or lemon lime soft drink)

To make blackberry lime syrup, place blackberries in medium bowl. Add lime juice and sugar, and mash. Let sit for 10 minutes. Strain mixture through fine-mesh sieve to remove seeds.

To make drinks, fill 4 large glasses with ice cubes and pour 1/4 cup (60 mL) blackberry lime syrup in each glass. Top up glasses with club soda.

Pictured on page 33.

Homemade Raspberry Lemonade

Serves 6

In the heat of summer, lemonade is the best thing to quench my thirst. Water may be the best for dehydration, but the tastiest alternative for me is my Homemade Raspberry Lemonade, made with fresh raspberries that we pick ourselves. Let's just say, there is nothing like it. This drink will quickly become a summer favourite in your home; we always keep a big jug of it in the fridge, nicely chilled with some ice.

> 1 cup (250 mL) granulated sugar
> 1 cup (250 mL) water
>
> 3/4 cup (175 mL) fresh raspberries
>
> 1 cup (250 mL) fresh-squeezed lemon juice (about 8 lemons)
> 5 cups (1.25 L) water

In medium saucepan, combine sugar with first amount of water. Heat over medium, stirring occasionally, until sugar has completely dissolved. Transfer to 2 quart (2 L) pitcher and set aside to cool.

Place raspberries in blender or food processor and process until smooth. Push raspberry purée through fine-mesh sieve to remove seeds. Add raspberry purée to pitcher.

Add lemon juice and second amount of water. Adjust amount of water to your taste, to achieve your perfect sweet/tart balance. Chill before serving.

Raspberry Watermelon Slush

Serves 4

Because slush is a drink that my daughters go crazy for during summer, I needed to find a recipe that I could make easily at home. It irked me to purchase mass-produced slush made with sugary syrup in machines at convenience stores. Raspberry Watermelon Slush is not only simple but also extremely tasty and relatively healthy. It is made with real fruit and is especially great when watermelons are available and affordable and when you have raspberries on hand, whether wild or cultivated. For the adults in the household, you could spruce it up for a party by adding some vodka or gin.

> 3 cups (750 mL) cubed watermelon
> 1 1/2 cups (375 mL) fresh raspberries
> 2/3 cup (150 mL) orange juice
> 2 Tbsp (30 mL) honey
> 1 Tbsp (15 mL) lime juice

Combine all 5 ingredients in blender and process until smooth. Pour mixture into freezer-safe container and freeze for minimum 5 hours. When ready to serve, remove from freezer, let stand at room temperature for 15 to 20 minutes, then cut into chunks and put back in blender. Pulse frozen chunks for 20 to 30 seconds. Pour slush into 4 tall glasses and serve immediately with a straw and a spoon.

Berry Spritzer

Serves 8

On top of its high levels of antioxidants, the blackberry is known for promoting alertness and increasing brain activity. The blackberry also carries with it a multitude of other soothing properties. Not only will it help reduce the risk of cancer, but it can also promote tightening of tissue as a natural way of making your skin appear younger. So, if someone asks you what you are doing as you sip on your Berry Spritzer, you can just explain that you are working toward a more youthful appearance and that your drink is helping you out.

> 1/4 cup (60 mL) granulated sugar
> 1/4 cup (60 mL) water
>
> 3 cups (750 mL) fresh blackberries
> 1 cup (250 mL) fresh raspberries
> 1/2 cup (125 mL) white grape juice
> 1/3 cup (75 mL) lemon juice
>
> ice cubes, for serving
> 1 1/2 cups (375 mL) sparkling white wine, chilled

In small saucepan over medium, bring sugar and water to a boil and then simmer, stirring, until sugar has dissolved. Remove from heat and set aside to cool.

Place cooled sugar mixture, blackberries, raspberries, grape juice and lemon juice in blender and process until smooth. Strain mixture through fine-mesh sieve, discarding solids.

Fill 8 glasses with ice and divide berry mixture evenly between glasses; top each one with sparkling wine and serve.

Strawberry Banana Pancakes

Serves 4

When it comes to cultivated strawberries in Canada, there are two main varieties of berries: the June-bearing variety and the day-neutral variety. June-bearing strawberries, as the name implies, will produce one crop of berries per year, which is usually in the month of June, depending on the climate. Day-neutral strawberry plants may produce a berry crop right through the summer, often until September or even November in some warm regions. The plants of the day-neutral variety tend to be smaller than their June-bearing cousins because most of their energy goes into berry production.

1 egg, lightly beaten
2/3 cup (150 mL) milk
2 Tbsp (30 mL) vegetable oil
1 x 6.4 oz (180 g) pouch banana nut muffin mix

1 cup (250 mL) vanilla yogurt
2 cups (500 mL) sliced fresh strawberries
2 Tbsp (30 mL) icing sugar

Spray large skillet with cooking spray and preheat to medium-high. Combine egg, milk, oil and muffin mix in mixing bowl; stir until smooth throughout: Pour about 1/4 cup (60 mL) batter for each pancake onto hot skillet. Cook pancakes for 1 to 2 minutes, until bubbles begin to form on surface. Flip and cook another 1 to 2 minutes, until golden brown.

Spoon yogurt and strawberries on half of each pancake and fold over. Sprinkle each pancake with icing sugar and serve.

Fish Tacos with Strawberry Cucumber Salsa (p. 96),
Strawberry Piña Colada (p. 110)

Strawberry Rhubarb Crisp (p. 99)

Strawberry Waffles

Serves 4

Cooking homemade Belgian-style waffles is something I only picked up in the last few years. We were on vacation in New England and spent the night in Manchester, New Hampshire. There, we discovered a Belgian waffle maker in our hotel, and it changed our breakfast menu forever. The ability to make our own Belgian style waffles at home was an exciting prospect. We returned from that trip and quickly purchased a high-end waffle maker, and we have never looked back. This strawberry waffle recipe is something we prepare at least once a month, and it brings me back to New England every time.

2 cups (500 mL) sliced fresh strawberries
2 Tbsp (30 mL) granulated sugar

2 1/2 cups (625 mL) flour
1 1/2 Tbsp (22 mL) granulated sugar
4 tsp (20 mL) baking powder
3/4 tsp (4 mL) salt

2 eggs, slightly beaten
2 1/4 cups (550 mL) milk
3/4 cup (175 mL) vegetable oil

Preheat waffle iron. In small bowl, combine strawberries and first amount of sugar. Mix and set aside.

In medium bowl, combine flour, second amount of sugar, baking powder and salt. Mix.

In separate bowl, whisk together eggs, milk and oil. Add egg mixture to flour mixture and stir until smooth. Pour about 1/3 cup (75 mL) batter at a time onto centre of waffle iron. Cook for 3 to 5 minutes, until lightly browned. Serve warm topped with strawberry mixture.

Strawberry Rhubarb Muffins

Makes 12 muffins

The beloved rhubarb makes a great partner to many of our berries and is a healthy plant in its own right. Rhubarb is a terrific source of vitamin C to promote a healthy immune system, and an excellent source of vitamin K, calcium, potassium and magnesium, all of which are essential nutrients. Rhubarb is also a great source of fibre and can help lower your cholesterol. On top of all the excellent nutritional qualities, rhubarb boasts a certain tartness that helps balance the sweetness of strawberries in many great recipes such as this one: Strawberry Rhubarb Muffins.

2 1/2 cups (625 mL) flour
1 cup (250 mL) brown sugar
1 tsp (5 mL) baking soda
1/2 tsp (2 mL) salt

1 egg
3/4 cup (175 mL) buttermilk
1/4 cup (60 mL) vegetable oil
1 cup (250 mL) sliced fresh strawberries
1 cup (250 mL) sliced fresh rhubarb

1/2 cup (125 mL) brown sugar
1/2 tsp (2 mL) ground cinnamon

Preheat oven to 375°F (190°C). In large bowl, combine flour, first amount of brown sugar, baking soda and salt. Mix.

In separate bowl, whisk together egg, buttermilk and oil. Add egg mixture to flour mixture, stirring just until combined. Fold in strawberries and rhubarb. Spoon batter into 12 greased muffin cups.

In small bowl, combine second amount of brown sugar and cinnamon. Sprinkle on top of each muffin. Bake for 18 to 20 minutes, until wooden pick inserted in centre of muffin comes out clean.

Creamy Strawberry Spread

Serves 8 to 10 as an appetizer

According to the Ontario Ministry of Agriculture and Food, the department that oversees the production of strawberries, most strawberries grown on Ontario farms are planted using the matted row method. In the first year, plants are planted in evenly spaced rows and then allowed to grow vegetatively, sending out runners and establishing daughter plants, eventually creating a matted tangle of strawberry plants—hence the name. The berries are harvested the second season for approximately two to four years after planting, after which the plants are allowed to die off over winter. The fields are then left fallow for a year. Strawberry plants are tender perennials, and almost everywhere in the province, they have to be protected in winter with straw mulch to prevent winter die-off.

> 1 x 8 oz (250 g) block cream cheese, softened
> 1/2 cup (125 mL) salad dressing (or mayonnaise)
> 1/2 cup (125 mL) mashed fresh (or frozen, thawed) strawberries
> 1/2 cup (125 mL) crumbled feta cheese
> 1/2 cup (125 mL) sliced almonds
> 1/2 cup (125 mL) diced red pepper
> 1/2 cup (125 mL) diced shallots

In medium bowl, combine all 7 ingredients and mix thoroughly. Serve with your choice of crackers.

Brie Cheese with Strawberry Preserves

Serves 6 to 8 as an appetizer

For many Canadians who grew up in the more rural regions of this country, the strawberry tea social was a big part of the community. Strawberry teas were often celebrated in a local church basement or community centre and usually featured light snacks such as cookies and beverages, as well as offering the star of the event—strawberries—prepared in several different ways. I can recall as a child our church's strawberry tea socials. It was a fun day to visit with neighbours and enjoy some terrific berries.

1/2 x 14 oz (397 g) package puff pastry, thawed

2 x 4 oz (125 g) brie cheese rounds
4 Tbsp (60 mL) strawberry preserves (see Note)

1 egg white

Preheat oven to 350°F (175°C). Roll out pastry to a 6 x 9 inch (15 x 23 cm) rectangle. Cut in half.

Place 1 round of brie on 1 half of pastry. Spread 2 Tbsp (30 mL) strawberry preserves on brie. Wrap with puff pastry and transfer onto greased cooking sheet, seam side down. Repeat with second round of brie.

Brush pastry with egg white and bake for about 30 minutes, until pastry is golden brown. Serve warm along with crackers as an appetizer.

Note: If you don't want to use store-bought preserves, you can make your own; Strawberry Preserves, page 108, will work well in this recipe.

Strawberry Steak Salad

Serves 4

You've heard of chicken salad, tuna salad and pasta salad, but steak salad? Beef and salad do not always go hand in hand, but once you've tried this delight, you will be a believer. There are two key elements to make steak a success. First, know how to properly cook a steak without overcooking it. Although this recipe calls for 4 minutes per side on medium-high, you may want to start on high for a quick sear on both sides to help lock in the flavour and tenderness before turning down the heat. Second, know how to cut "across the grain." The grain of a cooked steak will appear in thin parallel lines, and when you cut across the grain, the resulting piece of meat will be much more tender to chew because the long natural grains have been cut shorter.

3 Tbsp (45 mL) olive oil
3 Tbsp (45 mL) strawberry jam (see Note), warmed
2 Tbsp (30 mL) white balsamic vinegar
1/4 tsp (1 mL) pepper

1 x 12 oz (340 g) sirloin steak
salt and pepper

4 cups (1 L) spinach leaves
4 cups (1 L) romaine lettuce leaves
2/3 cup (150 mL) sliced fresh strawberries
2 green onions, chopped
1/2 cup (125 mL) walnut pieces, toasted
4 oz (125 g) goat cheese, crumbled

In small bowl, whisk together olive oil, strawberry jam, balsamic vinegar and pepper. Set aside.

Preheat grill to medium-high. Season steak with salt and pepper. Cook on greased grill for about 4 minutes per side, until desired doneness. Remove from grill and set aside to rest.

In large bowl, combine spinach, lettuce, strawberries and green onion. Add olive oil mixture and toss. Cut steak thinly across grain and add to salad. Top with walnuts and goat cheese, and serve.

Note: If you don't want to use store-bought jam, you can make your own; Easy No-cook Strawberry Freezer Jam, page 107, will work well in this recipe.

Braised Pork with Strawberry Whisky Sauce

Serves 6 to 8

Picking strawberries is of one the favourite early summer activities in my family. We love nothing more than heading off to the nearby strawberry farm with pails in our hands. Since the strawberry ripeness window is very small during June and July, it is important to act quickly and get picking while the picking is good. Most strawberry growers will plant several varieties, with varied ripening times. Once one field has ripened and been picked, the adjacent field planted with a slightly different strain hits prime ripeness, and so on. This staggered strategy is productive and cost-effective for growers, and is very much appreciated by consumers like us, who can't get enough of these wonderful red berries.

4 garlic cloves, minced
2 Tbsp (30 mL) Canadian whisky
2 Tbsp (30 mL) grated ginger root
1 Tbsp (15 mL) lemon juice
3/4 tsp (4 mL) salt
3/4 tsp (4 mL) pepper
1/4 tsp (1 mL) cayenne pepper
1 x 2 1/2 lbs (1.1 kg) pork loin or roast

2 Tbsp (30 mL) vegetable oil

2 garlic cloves, minced
3/4 cup (175 mL) mashed strawberries
1/4 cup (60 mL) brown sugar
1/4 cup (60 mL) Canadian whisky
3 Tbsp (45 mL) molasses

In large resealable plastic bag, combine first amounts of garlic and whisky, ginger, lemon juice, salt, pepper and cayenne. Add pork and gently shake bag to ensure marinade covers meat. Refrigerate for 4 to 24 hours.

Remove pork from bag and discard marinade. Heat oil in skillet over medium-high and brown all sides of meat. Transfer to medium casserole dish.

(continued on next page)

Preheat oven to 350°F (175°C). In small saucepan, combine second amount of garlic, strawberries, brown sugar, second amount of whisky and molasses. Bring to a gentle boil. Boil for 10 to 12 minutes, until slightly thickened. Pour over meat in casserole dish. Cover and cook in oven for 1 hour. Turn meat in sauce at least once. Cook uncovered for another 30 minutes. Let stand for 5 minutes before serving.

Summertime Strawberry Pork Chops

Serves 4

Anytime I get the opportunity to prepare a meal with pork chops is a good time, in my opinion. Pork is a healthy protein and provides a wide range of vitamins and minerals, particularly the B vitamins and the minerals zinc, iron, magnesium and phosphorus, all of which play a role in cellular and metabolic functions. Regardless of the berry they are served with, pork chops are a wise choice and always delicious, especially when prepared in the summertime.

2 tsp (10 mL) vegetable oil
4 pork chops, 3/4 inch (2 cm) thick

1/4 cup (60 mL) strawberry preserves (see Note)
1/4 cup (60 mL) apple cider vinegar
1 Tbsp (15 mL) prepared mustard

Heat oil in large skillet over medium-high and cook chops until brown on both sides. Reduce heat to low.

In small bowl, combine strawberry preserves, vinegar and mustard. Mix well. Pour sauce over chops. Cover and cook until sauce has thickened and meat has fully cooked, about 8 minutes. To serve, spoon sauce over each chop.

Note: If you don't want to use store-bought preserves, you can make your own; Strawberry Preserves, page 108, will work well in this recipe.

Fish Tacos with Strawberry Cucumber Salsa

Serves 4

Halibut are the largest flatfish in the world and are capable of reaching staggering weights of up to 300 kilograms. These fish grow extremely quickly and live for a very long time. Many specimens caught by fishermen are 20-plus years old, and one geezer halibut was even recorded at 55 years old. Halibut are highly sought after by anglers off the coast of British Columbia for their sporting quality and their taste. The International Pacific Halibut Commission is an organization established by the United States and Canada in 1923. It has worked to manage and conserve halibut waters shared by the two countries.

2 cups (500 mL) diced fresh strawberries
1 English cucumber, peeled and diced
1 jalapeño pepper, seeded and diced
2/3 cup (150 mL) diced shallots
2 Tbsp (30 mL) chopped fresh cilantro
2 Tbsp (30 mL) lemon juice

1/4 cup (60 mL) lemon juice
2 Tbsp (30 mL) olive oil
3 garlic cloves, minced
1/2 tsp (2 mL) salt
1/2 tsp (2 mL) pepper
1 lb (454 g) white fish fillets such as halibut, cod or sole

2 Tbsp (30 mL) vegetable oil

2 cups (500 mL) shredded cabbage
8 x 6 inch (15 cm) corn tortillas

For salsa, combine strawberries, cucumber, jalapeño, shallots, cilantro and first amount of lemon juice. Place mixture in airtight container and refrigerate until ready to use.

In small bowl, combine second amount of lemon juice, olive oil, garlic, salt and pepper; mix well and pour into shallow baking dish. Add fish and turn to coat. Allow to marinate in refrigerator for at least 30 minutes but no longer than 2 hours.

(continued on next page)

Strawberry

Remove fish from marinade and discard marinade. Heat oil in skillet and add fish. Cook for 6 to 10 minutes, flipping halfway, depending on thickness of fillets. When done, fish should flake easily with a fork but not be too dry. Slice fish into large pieces.

To assemble tacos, divide cabbage among tortillas. Add fish and top with a dollop of strawberry salsa. Serve.

Pictured on page 87.

Strawberry Chicken Strips

Serves 4

Chicken white meat is extremely popular in Canada for young and old alike, and it can be served in a variety of different ways. In this recipe, the naturally mild flavour of chicken pairs beautifully with the sweetness of strawberries, and the acidity and savouriness of the Russian dressing serves as the perfect accent point. Try pairing chicken with other berries—the possibilities are endless!

1 1/4 cups (300 mL) Russian dressing
1/4 cup (60 mL) strawberry jam (see Note)
3 Tbsp (45 mL) onion soup mix
2 Tbsp (30 mL) curry powder

1 lb (454 g) boneless, skinless chicken breast halves,
   cut into strips

In small bowl, combine Russian dressing, strawberry jam, onion soup mix and curry powder. Mix well.

Place chicken strips in medium baking dish and pour dressing mixture over top. Cover and refrigerate for 2 hours. Preheat oven to 350°F (175°C). Cook chicken for 20 to 30 minutes, until no longer pink in centre. Serve hot.

Note: If you don't want to use store-bought jam, you can make your own; Easy No-cook Strawberry Freezer Jam, page 107, will work well in this recipe.

Strawberry Fritters

Serves 6

The strawberry is perhaps the most popular of all berries in Canada. However, strawberries are extremely perishable with a narrow ripeness window. They are typically sold to consumers within two days of being picked and are often marketed on a "pick your own" basis. In fact, Canadian strawberry growers estimate that over 90 percent of all strawberry sales are from consumers picking their own. More and more, Canadians are going one step further and growing their own; the sale of strawberry plants is on the rise across the country.

1/4 cup (60 mL) butter, melted
1/4 cup (60 mL) half-and-half cream
3 eggs
1 cup (250 mL) flour
2 Tbsp (30 mL) brown sugar
1 tsp (5 mL) ground cinnamon
1/4 tsp (1 mL) salt
2 cups (500 mL) hulled fresh strawberries

4 cups (1 L) vegetable oil, for frying

In large bowl, whisk together butter, cream and eggs. Add flour, brown sugar, cinnamon and salt. Mix well. Gently fold in strawberries.

Heat oil in deep fryer to 360°F (180°C) or in large, heavy pot over medium-high. Carefully drop coated strawberries into hot oil one at a time, working in small batches. Fry for 5 minutes or until batter is golden brown, turning to brown all sides evenly. Remove cooked strawberry fritters and place on paper towels to drain.

 COOKING OIL TEMPERATURE

Cooking with oil can be a fun and tasty way to prepare several dishes, but finding the right cooking temperature can be difficult if you don't have a deep-fry thermometer. A simple technique called the "corn kernel pop" has served me well over the years. You simply drop one popcorn kernel into your oil when it is heating up. The kernel will pop between 350°F and 375°F (175°C and 190°C), which is the optimal oil temperature for most dishes. Once the kernel pops, simply take it out and start cooking.

Strawberry Rhubarb Crisp

Serves 6 to 8

If you live in eastern Canada's beautiful province of Nova Scotia, chances are that you have enjoyed more than a few strawberries over the years. Nova Scotians consume more strawberries per capita than any other region in Canada, and they have the strawberry farms to prove it. Consult Horticulture Nova Scotia for tips on picking and growing, recipes and anything else that you ever wanted to know about strawberries but didn't know who to ask. Strawberry growers in Nova Scotia take pride in the quality of their berries, and the consensus is that the local strawberries are a top-notch product at a very good price.

3 cups (750 mL) diced fresh (or frozen, thawed) rhubarb
1 1/2 cups (375 mL) sliced fresh strawberries
1/2 cup (125 mL) strawberry juice
3/4 cup (175 mL) granulated sugar
1 1/2 Tbsp (22 mL) minute tapioca

1/4 cup (60 mL) butter, melted
1 cup (250 mL) rolled oats
1/2 cup (125 mL) flour
3/4 cup (175 mL) brown sugar
1 tsp (5 mL) ground cinnamon

Preheat oven to 375°F (190°C). In 2 quart (2 L) baking dish, combine first 5 ingredients. Mix and let stand for 5 minutes.

In medium bowl, combine remaining 5 ingredients. Mix. Sprinkle over rhubarb mixture. Bake for about 25 minutes, until rhubarb is cooked.

Pictured on page 88.

Vanilla Bean Fruit Salad

Serves 8

The majority of home cooks opt for vanilla extract in a recipe rather than fresh vanilla bean; however, until you have tried fresh vanilla bean, you haven't unlocked the true potential of vanilla. When choosing the perfect vanilla bean for your recipe, look for a pod about 15 centimetres long with a dark brown skin that is moist to the touch and smells wonderful. Even when buying a rewrapped bean from a specialty shop or supermarket, the beautiful vanilla aroma should waft through. Try the vanilla bean; it will surprise you.

> 1 cup (250 mL) granulated sugar
> 1/2 cup (125mL) water
> 1 vanilla bean
> 1/2 tsp (2 mL) grated lime zest
>
> 5 cups (1. 25 L) cubed fresh pineapple
> 2 cups (500 mL) cubed cantaloupe
> 4 kiwi fruits, quartered
>
> 4 cups (1 L) stemmed and quartered fresh strawberries
> 1/2 cup (125 mL) sparkling white wine
> 12 fresh mint leaves, torn

Combine sugar and water in small saucepan. Split vanilla bean in half lengthwise. Scrape seeds from pod into sugar mixture, and add pod halves. Heat, stirring, on medium until sugar is dissolved, about 4 minutes. Increase heat to medium-high. Boil for about 5 minutes, without stirring, until slightly thickened. Remove from heat and stir in lime zest. Let stand for 10 minutes. Pour into extra-large bowl and chill, covered, for 1 hour. Remove and discard pod halves. Leave seeds in syrup.

Add pineapple, cantaloupe and kiwi to syrup. Stir until coated. Chill, covered, for at least 3 hours, stirring occasionally.

Stir in strawberries, wine and mint. Spoon into 8 dessert bowls and serve.

Strawberry

Traditional Strawberry Shortcake

Serves 6 to 8

As you enjoy your Traditional Strawberry Shortcake, consider this bit of trivia: according to the Guinness Book of World Records, the world's largest strawberry was grown by G. Anderson of Folkstone, Kent, in 1983. The massive berry tipped the scales at a staggering 231 grams (8.15 ounces). Most cultivated strawberries weigh approximately 5 to 10 grams and are nowhere near the size of that one grown in England.

1 1/2 lbs (680 g) fresh strawberries, sliced
3 Tbsp (45 mL) granulated sugar

2 cups (500 mL) flour
2 Tbsp (30 mL) granulated sugar
2 tsp (10 mL) baking powder
1/4 tsp (1 mL) baking soda
3/4 tsp (4 mL) salt
1 1/2 cups (375 mL) half-and-half cream

1 cup (250 mL) whipping cream

Place strawberries in bowl. Mix in first amount of sugar and let sit for about 30 minutes so juices come out.

Preheat oven to 400°F (200°C). In large bowl, combine flour, second amount of sugar, baking powder, baking soda and salt. Mix. Add cream and mix just until combined. Transfer batter to greased 8 x 8 inch (20 x 20 cm) baking pan. Bake for 18 to 20 minutes, until golden. Remove shortcake from pan and place on a wire rack to cool slightly.

Pour whipping cream into deep bowl. Beat until soft peaks form. To serve, cut cake into 6 or 8 pieces and split each piece in half horizontally. Spoon some strawberries with their juice onto each shortcake bottom half. Top with generous dollop of whipped cream and then the shortcake top half. Spoon more strawberries over top.

Chocolate-covered Strawberries

Serves 2 to 4

In ancient times, strawberries were associated with the Roman love goddess Venus. Even today, strawberries are often considered to be a romantic food, the "fruit of love," but why is that? It may have something to do with the strawberry's heart-shaped form, its sweetness and its passionate red colour. So the next time you plan a romantic dinner date, keep a bowl of fresh strawberries around. You never know what might happen.

1 lb (454 g) fresh strawberries, whole

2 cups (500 mL) milk chocolate chips
2 Tbsp (30 mL) shortening

Insert wooden picks into tops of strawberries.

In double boiler (see Note), heat chocolate chips and shortening, stirring occasionally, until melted and smooth. Holding them by wooden picks, dip strawberries, one at a time, into chocolate mixture. Place each strawberry on baking sheet lined with wax paper. Chill in refrigerator for 30 minutes.

Note: If you don't have or don't want to use a double boiler, heat your chocolate chips and shortening in microwave on high 1 minute at a time, stirring after every minute, until melted.

Strawberry Rhubarb Stew

Makes about 7 cups (1.75 L)

Rhubarb grows mostly in southern areas of our country. Whether picked wild or cultivated in your own garden, harvesting rhubarb at the proper time is key to success. Stalks should be about 30 centimetres in length and firm to the touch. If you leave rhubarb too long in summer, although thicker in size, it becomes very tough and not as enjoyable to eat. Harvest rhubarb selectively throughout the growing season; be sure to leave behind approximately one-third of the stocks to ensure good future growth.

> 4 cups (1 L) chopped fresh (or frozen, thawed) rhubarb
> 2 medium apples, peeled and cored, cut into 8 wedges each
> 2 cups (500 mL) dried apricots, halved
> 1 cup (250 mL) water
> 1/4 cup (60 mL) liquid honey
> 2 Tbsp (30 mL) minute tapioca
> 2 cinnamon sticks
>
> 3 cups (750 mL) stemmed and quartered fresh strawberries
> 1/2 tsp (2 mL) vanilla extract

Combine rhubarb, apples, apricots, water, honey, tapioca and cinnamon sticks in 3 1/2 to 4 quart (3.5 to 4 L) slow cooker. Cook, covered, on Low for 5 to 6 hours or on High for 2 1/2 to 3 hours. Remove and discard cinnamon sticks.

Add strawberries and vanilla. Stir. Cook, covered, on High for about 20 minutes, until strawberries are heated through. Serve warm or cold over ice cream, frozen yogurt or angel food cake.

Summertime Strawberry Barbecue Sauce

Makes 3 cups (750 mL)

For me, picking strawberries and fishing always went together. Every weekend during summer, I would stroll through the back fields to my fishing hole, a deep bend in Beaven Creek known as Fiddler's Elbow. It seemed that the best pike and bass fishing always coincided with strawberry season, so I made sure to reserve some extra space in my fishing tackle box to store the yummy strawberries I would stop and pick on the way back to the house. Over time, fishing at Fiddlers' Elbow became synonymous with great strawberry picking, and those outings when the fish weren't biting wouldn't seem that bad—a good feed of strawberries always eased the pain.

> 2 cups (500 mL) sliced fresh strawberries
> 1/3 cup (75 mL) strawberry jam (see Note)
> 1/2 cup (125 mL) ketchup
> 3 Tbsp (45 mL) chopped shallots
> 2 Tbsp (30 mL) honey
> 2 Tbsp (30 mL) lemon juice
> 2 Tbsp (30 mL) soy sauce
> 1 tsp (5 mL) hot sauce
> 1 tsp (5 mL) liquid smoke
> 1/4 tsp (1 mL) crushed red pepper
> 1/4 tsp (1 mL) salt

Place all 11 ingredients in blender or food processor and blend until smooth. Makes great barbecue sauce for both poultry and fish. Keeps well in refrigerator for up to 3 months.

Note: If you don't want to use store-bought jam, you can make your own; Easy No-cook Strawberry Freezer Jam, page 107, will work well in this recipe.

Strawberry

Lamb Chops with Blackcurrant Sauce (p. 114)

Grilled Brookies with Gooseberry Sauce (p. 115)

Strawberry Vinaigrette

Makes 1 cup (250 mL)

Recipes like this one are perfect for people who like sweet foods but cannot have a lot of sugar in their diets. Keep in mind, however, that honey changes things. Although completely natural, honey will affect a person's blood sugar levels as much as granulated sugar.

1 cup (250 mL) chopped fresh strawberries
1 Tbsp (15 mL) white wine vinegar
1 1/2 tsp (7 mL) balsamic vinegar
1 tsp (5 mL) liquid honey (optional)
1 tsp (5 mL) pepper

Process all 5 ingredients in blender until smooth. Serve over salad greens. Dressing keeps well in refrigerator for up to 3 months.

Easy No-cook Strawberry Freezer Jam

Makes about 6 cups (1.5 L)

Freezer jam is the king of all homemade jams in this country. There are several reasons for its popularity, with ease of preparation topping the list. Another reason strawberry freezer jam is so popular is the taste. It is simply delicious and, to me, is one of the most palatable of all homemade jams. No other jam retains the flavour of the original berry in quite the same way. Every time I take a bite of toast with this jam, it feels like eating fresh strawberries in the middle of strawberry season, even if it's mid-February.

2 cups (500 mL) crushed fresh strawberries
4 cups (1 L) granulated sugar
1 x 3 oz (85 mL) pouch liquid pectin
2 Tbsp (30 mL) lemon juice

Place strawberries in large bowl. Add sugar and stir. Let stand for 10 minutes. Add pectin and lemon juice and stir for 2 minutes. Pour mixture into clean jars or plastic containers, leaving enough room for the mixture to expand when it freezes. Let stand at room temperature for 24 hours or until set. Place containers or jars in freezer until ready to use.

Strawberry Preserves

Makes 4 cups (1 L)

Because so many of us enjoy eating berries at times of the year other than summer, summer is when we need to stock up on our supply. Freezing, drying and canning are three great ways to enjoy any type of berry throughout the year, but these Strawberry Preserves are my favourite way to enjoy strawberries on a cold, snowy day. If the winter blues are getting you down, anything that reminds you of summer is a good thing.

5 cups (1.25 L) small fresh strawberries, hulled
peel (with white pith) of 1/2 lemon
1 cup (250 mL) granulated sugar

Place strawberries in large, heavy pot. Add lemon peel. Fold in sugar. Let stand at room temperature, stirring occasionally, for 2 hours. The berries will release their juices and the sugar will dissolve.

Bring strawberry mixture to a gentle boil on medium. Cook, stirring occasionally, until berries are just tender, about 3 minutes. Remove from heat. Using slotted spoon, transfer strawberries to two 2 cup (500 mL) hot sterile jars. Return pot with berry juices and lemon peel to medium-high and boil for 1 to 2 minutes, until syrup thickens. Remove from heat and let syrup settle. Skim off any foam. Discard lemon peel. Pour hot syrup over strawberries. Wipe rims clean. Place hot metal lids on jars and screw on metal bands fingertip tight. Allow to cool, then refrigerate. Keeps well in refrigerator for up to 3 months.

Strawberry Banana Smoothie

Serves 2

The best smoothies, I find, are made with frozen berries because the added moisture brings just the right amount of liquid to your favourite drink. Frozen berries, whether strawberries, raspberries or blackberries, are also a great alternative to ice cubes. You don't have to water down your shake, smoothie or cocktail if you use berries as ice cubes. My family often freezes our berries after foraging and stores them in well-sealed, labelled containers. They will keep fresh for a year and more and are just as tasty thawed as they were when first picked.

> 1 cup (250 mL) any flavour yogurt
> 1 cup (250 mL) any flavour fruit juice
> 1 ripe banana
> 1 cup (250 mL) frozen strawberries, whole or sliced

Combine all 4 ingredients in blender, and process until smooth. Pour into 2 glasses, place a straw in each one and enjoy.

 tip SMOOTHIE POPS

Essentially, you can turn any smoothie into popsicles by simply pouring it into popsicle moulds from your local dollar store and freezing them. Smoothie popsicles are a great healthy alternative to store-bought popsicles, ice cream or other treats. On a hot day, your children will be thrilled to have one of these for breakfast! We like to pre-freeze several different varieties of frozen pops to cover every urge our kids might have. By replacing the sugary processed treats with natural ones, our children's health will benefit greatly in the long run.

Strawberry Jalapeño Shooters

Makes 12 shooters

Strawberry Jalapeño Shooters are a great summertime beverage that carry with them quite the kick. I discovered that the spicy nip of the popular appetizer known as the jalapeño popper carries over nicely into a shooter. But be prepared for some serious heartburn with this drink. These shooters go down best after you have a little food in your stomach. Serve them after your guests have had time to munch on a few appetizers first.

> 1 lime
> 2 cups (500 mL) halved fresh strawberries
> 1/4 cup (60 mL) granulated sugar
> 1/4 cup (60 mL) water
> 1/4 jalapeño pepper, seeded
> 1 1/4 cups (300 mL) lemon vodka
>
> coarse sanding sugar, for rimming glasses

Zest and juice lime. Place strawberries, lime juice, lime zest, sugar, water and jalapeño in medium saucepan and bring mixture to a simmer. Cook for 5 minutes. Transfer to blender and pulse once (only once is important). Strain mixture through fine-mesh sieve. Add vodka to strained mixture and chill in refrigerator.

To serve, moisten rims of shot glasses. Dip rims into coarse sugar. Pour chilled strawberry mixture into prepared glasses and serve immediately.

Strawberry Piña Colada

Serves 4

Any time I taste that wonderful combination that is pineapple and coconut, it brings me back to vacations in the Caribbean. Hot, lazy days sitting in the sun, sipping on a cool, refreshing beverage… It's a nice way to spend a couple of weeks. Trying to replicate a Jamaican piña colada when I returned home was always a quest of mine; however, in the case of my favourite tropical drink, I discovered that strawberry is the perfect additive for a delightful Canadian piña colada.

(continued on next page)

4 cups (1 L) ice cubes
1 cup (250 mL) white rum
2 cups (500 mL) pineapple juice
1/2 cup (125 mL) coconut cream
24 large fresh strawberries, stemmed and quartered
4 pineapple wedges

Place ice cubes in blender and pulse until ice is crushed. Add rum, pineapple juice, coconut cream and strawberries. Blend until smooth. Pour into 4 tall glasses and garnish each one with a pineapple wedge.

Pictured on page 87.

Strawberry Cotton Candy Spritzer

Serves 1

For a beverage that will bring you back to your youth, one that is reminiscent of those country fairs that rolled through town each summer, the Strawberry Cotton Candy Spritzer fits the bill. It also has a tasty little kick. My wife and I often make these drinks at the trailer as a treat if we have guests over during the heat of August, and the sweetness appeals to most everyone. Give it a try; if you drink the right amount, you'll actually feel like you're your back at the fair again.

4 to 6 fresh strawberries, crushed
1 cup (250 mL) cotton candy
1/4 cup (60 mL) cotton candy liqueur
1 cup (250 mL) lemon lime soft drink
3 ice cubes, crushed

Place strawberries in tall glass. Add cotton candy. Pour in cotton candy liqueur. Pour in lemon lime soft drink. Add crushed ice and stir. Serve immediately.

Canadian Currant Buns

Makes 12 buns

Both red and black currants grow naturally in Canada, and they are also cultivated to some extent. Although very similar, redcurrant tends to be more tart, and blackcurrant is described as slightly more savoury. Currant picking is very popular in southern and southwestern Ontario, where there are many "pick your own" currant farms. My uncle, David Jones, has a property in Jordan Station, Ontario, where he annually forages for both red and black currants that grow wild on his land to use in a variety of recipes. Each summer, my parents are pleased when Uncle David's currant gift basket arrives, and the Canadian Currant Bun making can begin.

2 cups (500 mL) flour
1/2 tsp (2 mL) baking soda
1/2 tsp (2 mL) salt
1 Tbsp (15 mL) granulated sugar
1/4 cup (60 mL) shortening
3/4 cup (175 mL) buttermilk

butter, softened, for spreading over dough
1/4 cup (60 mL) granulated sugar
1/2 tsp (2 mL) ground cinnamon
1/4 cup (60 mL) chopped fresh black or red currants

Preheat oven to 375°F (190°C). In large bowl, combine flour, baking soda, salt and first amount of sugar. Cut in shortening. Add just enough buttermilk to produce stiff dough. Turn dough onto floured board. Knead slightly. Roll into rectangle about 1/4 inch (6 mm) thick.

Spread dough with butter. Sprinkle with second amount of sugar, cinnamon and currants. Roll up as you would a jelly roll. Cut into slices 3/4 inch (2 cm) thick. Place cut side down on baking sheet. Bake for 20 minutes.

Cocktail Wienies with Redcurrant Sauce

Serves 12 as an appetizer

If you are fortunate enough to have redcurrants growing nearby, you are among the chosen few in this country. With a somewhat narrow and picky growing season, these red gems are not nearly as ubiquitous as wild raspberries or blueberries. What they lack in availability, they make up for in character; they are one of the tastiest and most interesting of all berries. They are bright and tart like cranberries, but with a slightly sweeter flavour. Those who do have redcurrants growing on or near their property may find that some bushes produce better than others. When picking, focus on the largest berries for the best results.

> 1/2 cup (125 mL) redcurrant jelly (see Note)
> 1/2 cup (125 mL) Dijon mustard
> 2 Tbsp (30 mL) ketchup
> 1/4 cup (60 mL) brown sugar
> 2 cups (500 mL) cocktail wieners

In small bowl, combine redcurrant jelly, Dijon mustard, ketchup and brown sugar and mix well. Place cocktail wieners in slow cooker. Pour sauce over wieners. Simmer on High for at least 2 hours before serving with cocktail picks.

Note: If you don't want to use store-bought jelly, you can make your own; Redcurrant Jelly, page 120, will work well in this recipe.

Lamb Chops with Blackcurrant Sauce

Serves 4

Blackcurrant, as with its red cousin, is one of the underused berries in this country. It is a little known fact that the term "blackcurrant" is actually used in describing flavours and flavour palates of different recipes. The liqueur Cassis (French for "currant") and the table wine Cabernet Sauvignon are often described as having the intensity of blackcurrant. It is an interesting berry with a taste unlike any other. Experts describe the flavour as earthy, similar to gooseberry and passion fruit, with a hint of raspberry, and a floral aromatic note of roses and carnations, making it one classy berry indeed.

8 lamb chops, 1 1/4 inch (3 cm) thick, trimmed of most fat
salt and pepper
3 Tbsp (45 mL) olive oil

2 shallots, finely diced
1 large celery rib, finely diced
1 large parsnip, finely diced
1 large carrot, finely diced
2 cups (500 mL) dry red wine
4 cups (1 L) prepared chicken broth

8 black peppercorns
1/2 cup (125 mL) fresh blackcurrants
2 Tbsp (30 mL) butter
2 Tbsp (30 mL) fresh tarragon, finely diced

Preheat oven to 400°F (200°C). Season lamb chops on both sides with salt and pepper. In medium roasting pan set on two stove burners, heat olive oil over high. Add chops and cook for 2 to 3 minutes, until browned on one side. Turn chops over and transfer pan to oven. Cook for 5 to 6 minutes for medium-rare. Transfer 2 Tbsp (30 mL) liquid in roasting pan to medium saucepan. Cover roasting pan and allow meat to rest.

Place saucepan with reserved liquid on burner over high. Add shallots, celery, parsnip and carrot, and cook for about 10 minutes, until golden brown. Add wine and cook until completely reduced. Add chicken broth and cook until reduced to 1 cup (250 mL).

(continued on next page)

Strain sauce into small saucepan, reserving vegetables. Add peppercorns and currants to sauce and bring to a simmer over medium. Whisk in butter and tarragon. Serve chops drizzled with sauce and alongside vegetables.

Pictured on page 105.

Grilled Brookies with Gooseberry Sauce

Serves 4

Gooseberries and currants are closely related berries in the genus *Ribes,* and the gooseberry and blackcurrant have even ben crossed to produce a hybrid called the jostaberry. Although closely related, gooseberries and currants may be differentiated by closer inspection of their canes, and by the berries themselves. The cane of the gooseberry produces a spine or spike at each leaf node and produces berries either singly or in clumps of two to three. The currant cane, on the other hand, has no spines at all and bears small clusters of fruit in bunches. A mature gooseberry or currant shrub can produce up to four litres of fruit annually.

4 x 6 oz (170 g) brook trout fillets
salt and pepper

1 lb (454 g) fresh gooseberries, topped and tailed
1/2 cup (125 mL) dry white wine
1 egg
1/8 tsp (0.5 mL) ground ginger

Preheat grill to medium-high. Season trout with salt and pepper. Place, skin side down, on greased grill; close lid and cook for 8 to 10 minutes, until fillet flakes easily with a fork.

For sauce, place gooseberries in medium saucepan. Add wine and simmer for about 5 minutes, until fruit is very tender. Remove from heat. Beat in egg with a wooden spoon and cook for another 5 minutes. Season with ginger. Serve warm with cooked trout.

Pictured on page 106.

Blackcurrant Pound Cake with Tea Glaze

Serves 8 to 10

The beautiful blackcurrant, although not one of the most prevalent berries found in Canada, is of huge importance in Europe and New Zealand, the world's largest producers of this special berry. In June 2014, the International Blackcurrant Association held a conference in Poland that included 20 of the top growers in Europe and New Zealand. Blackcurrants account for one of the most important berry crops worldwide, and the conference allowed growers to discuss all aspects of the blackcurrant industry.

2/3 cup (150 mL) boiling water
6 blackcurrant tea bags
1/3 cup (75 mL) sour cream

3 cups (750 mL) flour
1 tsp (5 mL) baking powder
1/4 tsp (1 mL) salt

2 cups (500 mL) granulated sugar
3/4 cup (175 mL) butter, softened
3 eggs
2 tsp (10 mL) vanilla extract
1 cup (250 mL) fresh blackcurrants

1/4 cup (60 mL) boiling water
1 blackcurrant tea bag
3/4 cup (175 mL) icing sugar
1/2 tsp (2 mL) lemon juice

Pour first amount of boiling water over 6 tea bags in small bowl; steep for 10 minutes. Remove and discard tea bags; cool tea to room temperature. Stir in sour cream.

Preheat oven to 350°F (175°C). In large bowl, combine flour, baking powder and salt. Mix.

In another large bowl, combine granulated sugar and butter. Beat with electric mixer at medium speed until well blended. Beat in eggs, 1 at a time. Mix in vanilla. Add flour mixture and brewed tea mixture alternately

(continued on next page)

to sugar mixture, beginning and ending with flour mixture; mix after each addition. Fold in blackcurrants. Spoon batter into greased 10 inch (25 cm) tube pan. Bake for 1 hour and 10 minutes, or until wooden pick inserted in centre comes out clean. Cool in pan on wire rack for 15 minutes. Remove cake from pan and allow to cool completely on wire rack.

To prepare glaze, pour second amount of boiling water over 1 tea bag in small bowl; steep for 5 minutes. Remove and discard tea bag. Whisk in icing sugar and lemon juice. Drizzle over cake.

Gooseberry Custard

Serves 4 to 6

The gooseberry has been found to contain a multitude of health properties. It can improve eyesight by treating conjunctivitis, as well as alleviate glaucoma. The antibacterial and stringent qualities of gooseberry help fight infection. Gooseberry has also been found to help with diabetes management and assist in preventing heart disease. Native peoples historically used gooseberries to treat cardiovascular disease and other heart illnesses—gooseberries actually help strengthen heart muscles so that the heart will pump blood more freely through the body. This berry is basically an all-natural cure-all.

> 2 lbs (900 g) fresh gooseberries, topped and tailed
> 4 cups (1 L) water
>
> 2 Tbsp (30 mL) butter
> 1 Tbsp (15 mL) lemon juice
> 1 cup (250 mL) granulated sugar
>
> 1 cup (250 mL) half-and-half cream
> 4 eggs

In large saucepan over medium-high, combine gooseberries and water. Simmer for 7 to 10 minutes, until berries are soft. Drain water and gently mash berries.

Stir in butter, lemon juice and sugar and return to medium heat. Cook, stirring gently, for about 5 minutes, until sugar is completely dissolved.

In medium bowl, whisk cream and eggs together. Add cream mixture to gooseberry mixture. Cook on low, stirring constantly, for about 5 minutes, until thickened. Pour into serving bowl and chill for 2 hours, or until set.

Gooseberry Chutney

Makes 6 cups (1.5 L)

The glorious gooseberry, native to Europe and North Africa, has spread around the world. In England prior to World War I, gooseberry growing competitions were all the rage and continue to this day, though on a reduced scale. Gooseberries are harvested extensively for their goodness in many countries in Europe and also in New Zealand. In Hungary and Scandinavia, gooseberries are used in everything from soup to wine. Regardless of where they grow, these little green gems create quite a stir.

> 1 lb (454 g) fresh gooseberries, topped and tailed
> 1/2 cup (125 mL) water
> 1 large onion, diced
> 2 cups (500 mL) white wine vinegar
>
> 2 cups (500 mL) brown sugar
> 2 Tbsp (30 mL) salt
> 1 Tbsp (15 mL) ground ginger
> 1/2 tsp (2 mL) cayenne pepper

In large saucepan over medium-high, combine gooseberries and water. Cook for 7 to 10 minutes, until gooseberries have softened. Add onion and vinegar and continue cooking for 10 minutes.

Stir in brown sugar, salt, ginger and cayenne pepper. Boil gently for about 5 minutes, until some liquid has evaporated and mixture has slightly thickened. Remove from heat and allow to cool for 10 minutes. Ladle into hot sterile jars. Wipe rims clean. Place hot metal lids on jars and screw on metal bands fingertip tight. Store jars in refrigerator for up to 6 months.

Gooseberry and Pineapple Preserve

Makes 16 cups (4 L)

The wild gooseberry can be found across much of the Canadian boreal forest growing mostly as a riparian species, along creeks, riverbanks and lake shores. Aboriginal communities made extensive use of wild gooseberries, often mixing them with sweet corn and also grinding the plant roots for medicinal purposes. Gooseberries are now cultivated to a limited extent in Canada and seem to do well in our cooler climate. However, growers must be vigilant, as the gooseberry plant can serve as an alternative host for the nasty white pine blister rust, a fungal disease affecting pine trees in some areas.

3 lbs (1.6 kg) fresh gooseberries, topped and tailed
1 cup (250 mL) water
2 cups (500 mL) shredded fresh pineapple

2 cups (500 mL) seedless raisins
4 cups (1 L) granulated sugar
2 cups (500 mL) chopped pecans

In large saucepan, boil gooseberries in water for 10 to 12 minutes, until they burst. Add pineapple and cook for 10 minutes, stirring constantly.

Stir in raisins and sugar. Simmer, uncovered, for about 15 minutes, stirring frequently. Stir in pecans. Cook for about 5 minutes to heat them through. Ladle into hot sterile jars. Wipe rims clean. Place hot metal lids on jars and screw on metal bands fingertip tight. Store jars in refrigerator for up to 6 months.

Redcurrant Jelly

Makes 8 cups (2 L)

Because the ravishing redcurrant was not a berry native to the region that I grew up in, it was always a treat when my uncle would bring us a fresh batch of redcurrant jelly made from berries he harvested on his property in southwestern Ontario. Our fridge usually contained only raspberry and strawberry jam, so the redcurrant jelly was a nice addition to the breakfast table. Although currants would never completely replace the strawberries and raspberries we loved, we did appreciate the different taste that this more exotic berry offered.

4 lbs (1.8 kg) fresh redcurrants
1 cup (250 mL) water

6 cups (1.5 L) granulated sugar
3 oz (85 mL) liquid pectin

Crush currants and place in large saucepan. Add water and bring to a boil over medium-high. Reduce heat and simmer for 10 minutes. Strain fruit through jelly cloth or cheesecloth. Do not squeeze bag. Measure 5 cups (1.25 L) juice back into saucepan.

Stir in sugar. Bring to a rapid boil over high, and immediately stir in pectin. Return to a full rolling boil for 30 seconds. Remove from heat. Skim off any foam. Carefully ladle into hot sterile jars, leaving 1/4 inch (6 mm) headspace. Wipe rims clean. Place hot metal lids on jars and screw on metal bands fingertip tight. Process in boiling water bath for 10 minutes. Remove jars and allow to cool. Check that jars have sealed properly. Store sealed jars in a cool, dark place.

Saskatoon Berry Muffins with Topping

Makes 12 muffins

The saskatoon berry has long been steeped in tradition and played an important historical role for the aboriginals of this country. The name is from the Cree language. The leaves of the saskatoon bush were dried and used to make tea. The berries themselves, dark purple, delightful gems, were a diet staple. They were a common ingredient in pemmican and were also mashed and dried into a bricklike substance that would keep for long periods of time. Pieces of the brick were then chipped away as required and added as flavour to various dishes.

1 1/2 cups (375 mL) flour
3/4 cup (175 mL) granulated sugar
2 tsp (10 mL) baking powder
1/2 tsp (2 mL) salt

1 egg
1/3 cup (75 mL) vegetable oil
1/2 cup (125 mL) milk
1 tsp (5 mL) vanilla extract
1 cup (250 mL) fresh or frozen saskatoon berries

2/3 cup (150 mL) brown sugar
1/4 cup (60 mL) flour
1/4 tsp (1 mL) ground cinnamon
2 Tbsp (30 mL) butter

Preheat oven to 350°F (175°C). In large bowl, combine first amount of flour, granulated sugar, baking powder and salt. Mix.

In medium bowl, whisk together egg, oil, milk and vanilla. Add to flour mixture and stir just until combined. Fold in berries. Spoon batter into 12 greased muffin cups.

In small bowl, combine brown sugar, second amount of flour, cinnamon and butter. Mix with fork until crumbly. Sprinkle onto each muffin. Bake for 22 to 25 minutes, until wooden pick inserted into centre of muffin comes out clean.

Pictured on page 123.

ıskatoon Berry Lemon Bread

Makes 1 loaf

Saskatoon berries are a variety of berry that, sadly, I do not have the opportunity to enjoy as often as I would like. Native to the Canadian prairies, saskatoons are often mistaken for blueberries even though there are differences in both taste and appearance. They are more tart and a bit less juicy than the wild blueberry. They can be used in a plethora of recipes and are popular in pies and other recipes offering a distinct Canadian flavouring, earthy and somewhat sweet. Residents of Saskatchewan are proud to call the saskatoon their own.

1 1/2 cups (375 mL) flour
1 tsp (5 mL) baking powder
1/4 tsp (1 mL) salt
grated zest of 1 lemon

1/2 cup (125 mL) shortening
1 cup (250 mL) granulated sugar
2 eggs
1/2 cup (125 mL) milk
1/2 cup (125 mL) fresh or frozen saskatoon berries

1/4 cup (60 mL) lemon juice
1/4 cup (60 mL) granulated sugar

Preheat oven to 350°F (175°C). In medium bowl, combine flour, baking powder, salt and lemon zest. Mix.

In another medium bowl, mix shortening with first amount of sugar. Beat in eggs. Stir in milk. Add flour mixture and mix well. Gently fold in saskatoons. Pour into greased loaf pan. Bake for 45 minutes, or until wooden pick inserted in centre comes out clean. Set on wire rack to cool.

In small bowl, mix lemon juice and second amount of sugar. Drizzle over cooled loaf.

Saskatoon Berry

Saskatoon Berry Muffins with Topping (p. 121), Mulberry Muffins (p. 146),
Any Berry Milkshake (p. 170)

Cherry Upside-down Cake (p. 134)

Saskatoon Berry Bread Pudding

Serves 6 to 8

When visiting any of the Prairie Provinces, I suggest that you indulge in the region's tastiest trademark, the saskatoon berry. Various commercial saskatoon farms in Alberta, Saskatchewan and Manitoba feature both U-Pick and pre-picked options to obtain fresh berries, and several retail outlets boast a variety of jams, syrups, ciders and saskatoon berry gift baskets. A trip to the prairies would not be complete without experiencing the saskatoon.

2 cups (500 mL) cubed bread
1 cup (250 mL) fresh or frozen saskatoon berries

3 cups (750 mL) milk
2 Tbsp (30 mL) butter
2 tsp (10 mL) vanilla extract
1/8 tsp (0.5 mL) salt

4 eggs
1/2 cup (125 mL) granulated sugar

Preheat oven to 350°F (175°C). Lightly grease a 9 x 13 inch (23 x 33 cm) baking dish. Spread bread pieces evenly in baking dish. Spread saskatoons evenly over bread.

In microwave-safe bowl, heat milk in microwave for 3 minutes, stirring occasionally, until hot, then stir in butter so that it melts. Stir in vanilla and salt.

In medium bowl, lightly beat eggs. Add sugar and continue to beat. Whisk milk mixture into egg mixture. Slowly pour combined mixture over bread and berries. Bake for 35 minutes, or until centre of pudding is set. Serve warm or at room temperature.

Saskatoon Berry Pie

Serves 8

The Saskatchewan-based company Prairie Berries is one of several businesses catering to the province's most famous berry and offers this super fruit to customers around the world via their web-based store. Their products—from fruit syrups, toppings and pie fillings to dried and chocolate-coated saskatoon berries—feature no preservatives, additives or artificial colours or flavours. Prairie Berries invites you to indulge in the outrageously good flavours of the award-winning products they offer.

> **pastry for 2 crust, 9 inch (23 cm) pie**
>
> **4 cups (1 L) fresh saskatoon berries**
> **2 Tbsp (30 mL) water**
> **1 Tbsp (15 mL) lemon juice**
> **2/3 cup (150 mL) granulated sugar**
> **1/4 cup (60 mL) minute tapioca**

Preheat oven to 350°F (175°C). Line pie dish with pastry. Save remaining pastry for top crust.

Combine saskatoons, water and lemon juice in medium saucepan over medium. Simmer, stirring occasionally, for about 10 minutes (do not boil). Stir in sugar and tapioca. Pour mixture into unbaked pie shell. Cover with remaining pastry and press gently around edge to seal. Trim pastry edges and poke some holes randomly in top for steam vents. Bake for about 45 minutes, until juice is bubbly and crust is golden brown.

Saskatoon Berry Frozen Yogurt

Serves 4

The Saskatoon Berry Council of Canada (SBCC) is a not-for-profit organization that represents the Canadian saskatoon berry industry. The goal of the SBCC is to stimulate industry growth. The saskatoon berry industry in Canada is made up of nearly 1000 farms with a combined value of over $8 million for the Canadian economy. The greatest hurdle for market expansion is that saskatoons are relatively unknown in other parts of Canada and the world. The SBCC hopes to help spread the word of their product through education and awareness campaigns.

1 cup (250 mL) plain yogurt

1 cup (250 mL) fresh (or frozen, thawed) saskatoon berries
1 Tbsp (15 mL) honey
1 tsp (5 mL) vanilla extract

Place yogurt in freezer for 3 to 4 hours, until almost frozen.

Combine saskatoons, honey and vanilla in blender and process until smooth. Break yogurt into small chunks and add to blender. Process on low speed just until smooth. Transfer mixture to container; cover and freeze for another 3 to 4 hours, until frozen. Before serving, break frozen mixture into small chunks and transfer to blender. Blend at low speed just until smooth. Serve immediately.

Savoury Saskatoon Berry Sauce

Makes 2 cups (500 mL)

Saskatoon berries are known by a few different names, mostly according to geography. For much of our country, these little gems are commonly called saskatoons or occasionally shadberries; however, for our neighbours to the south, it's a different story. In August, 2014, the saskatoon berry was officially changed to juneberry in Michigan and Minnesota. Based on research at Cornell University, it was determined that American berry consumers loved the taste of saskatoons but were not as taken with the name. The Canadian Saskatoon Berry Council was asked to re-label saskatoons as juneberries for shipments to certain states. Western states such as Washington and California had already grown accustomed to the name saskatoon, so distribution of the berries to those areas did not require a name change.

2 cups (500 mL) fresh or frozen saskatoon berries
1 cup (250 mL) water
1 1/2 tsp (7 mL) beef bouillon powder
1/2 tsp (2 mL) minced garlic
1 Tbsp (15 mL) green peppercorns
1/2 tsp (2 mL) dried thyme
1/2 tsp (2 mL) dried rosemary

2 Tbsp (30 mL) cornstarch
1/2 cup (125 mL) granulated sugar
2 Tbsp (30 mL) lemon juice

Combine saskatoons and water in saucepan over medium-high and bring to a boil. Add bouillon powder, garlic, peppercorns, thyme and rosemary. Stir. Reduce heat and simmer for 5 minutes.

Mix cornstarch and sugar together and add gradually to saskatoon mixture. Boil until clear and thick. Stir in lemon juice and remove from heat. Serve over beef roast, pork tenderloin or chicken breast. It will keep well in a sealed container in the refrigerator for up to 3 months.

Saskatoon Berry Jam

Makes 6 cups (1.5 L)

Although this country's beloved saskatoon berries may appear similar to blueberries, they are quite different when it comes to processing. The skin of a saskatoon is much thicker than that of a blueberry, so you must take special measures when using saskatoons in a recipe. According to the SBCC, because the berries don't release their natural juices easily when baking, water should be added to assist in a natural leaching. When boiling saskatoons for jam, be careful not to overcook them, and always use a high grind setting on your blender to properly break down the skin and seeds of these hearty berries. Saskatoons are indeed delicious but also require a bit more care during processing.

> 8 cups (2 L) fresh saskatoon berries
> 1/4 cup (60 mL) lemon juice
> 1/2 tsp (2 mL) butter
> 1 x 2 oz (57 g) box pectin crystals
> 5 cups (1.25 L) granulated sugar

Crush saskatoons and place in large saucepan. Add lemon juice and butter (butter helps to reduce foaming). Stir. Add pectin. Bring to a boil over high, stirring constantly. Add sugar and stir until all sugar is dissolved. Continue stirring. Return mixture to a full rolling boil and boil hard for 1 minute, stirring constantly. Remove from heat. Skim off any foam. Stir and skim for 5 minutes to cool slightly. Carefully ladle mixture into hot sterile jars, leaving 1/4 inch (6 mm) headspace. Wipe rims clean. Place hot metal lids on jars and screw on metal bands fingertip tight. Process in boiling water bath for 15 minutes. Remove jars and allow to cool. Check that jars have sealed properly. Store sealed jars in a cool, dark place.

Cherry Coconut Muffins

Makes 12 muffins

Not only do these delicious muffins serve as a terrific dairy-free alternative to most other muffin recipes, but the flavour combination of coconut and cherry is also about as perfect a marriage as one could ask for. The inherent tartness of the sour cherries combined with natural flavour of coconut and coconut milk strike the perfect balance in these muffins. They are also a nice change from the ubiquitous blueberry and cranberry muffins you see on an almost daily basis. Easy to prepare and even easier to eat, these great muffins are sure to please.

3 cups (750 mL) flour
1 cup (250 mL) granulated sugar
1 cup (250 mL) medium unsweetened coconut
1 Tbsp (15 mL) baking powder
1/2 tsp (2 mL) salt

1 egg
1 x 14 oz (398 mL) can light coconut milk
1 Tbsp (15 mL) grated lemon zest
1 cup (250 mL) pitted and chopped sour cherries

Preheat oven to 375°F (190°C). In large bowl, combine flour, sugar, coconut, baking powder and salt. Mix.

In medium bowl, whisk together egg, coconut milk and lemon zest. Add to flour mixture and stir just until combined. Fold in cherries. Spoon batter into 12 greased muffin cups. Bake for 20 to 25 minutes, until wooden pick inserted in centre of muffin comes out clean.

130

Cherry

Cherry Pork Chops

Serves 4

Although not technically a berry, the cherry is often lumped in with other berries found in North America. Grown by some commercial growers in Canada, the cherry is an inherently sweet and unique food item that goes well in many recipes. I will never forget camping in British Columbia's beautiful Okanagan Valley, where several fruits, including berries and the most delicious cherries I've ever eaten, are grown each summer. Sitting on the shores of Okanagan Lake munching on a bag of fresh-picked cherries from one of the local orchards is something I will not soon forget. I remember wondering if the famous lake monster Ogopogo might swim by to snatch up the pits I was spitting into the water. But alas, the mysterious creature never showed up.

2 Tbsp (30 mL) vegetable oil
4 x 8 oz (250 g) pork chops
salt and pepper

1/4 cup (60 mL) butter
3/4 cup (175 mL) sliced shallots
1 cup (250 mL) pitted and halved fresh sweet cherries
1/4 cup (60 mL) prepared beef broth

Preheat oven to 350°F (175°C). Heat oil in skillet over medium-high. Season pork chops with salt and pepper, and cook for 2 to 4 minutes per side, until brown on both sides. Transfer chops to baking sheet. Cook in oven for 18 to 20 minutes, until no longer pink in centre.

Melt butter in skillet over medium. Add shallots and cherries and cook for 2 to 3 minutes, until shallots have started to soften. Stir in beef broth and bring to a slow simmer until sauce has reduced and thickened. Pour sauce over pork chops and serve.

Cherry Chicken Wraps

Serves 6

For anyone wanting to grow a cherry tree, there are a few useful tips to consider before you get started. First off, northern residents are warned that the cold climate and short growing season will stunt the growth of cherry trees, and they may never bear fruit. Residents of southern Canada are encouraged to spray new cherry trees with a dormant oil to protect against pests, and to stake newly planted trees to prevent wind damage. Prune cherry trees regularly, removing dead or dwindling branches. Birds and small mammals will target cherries, so you may want to install a predator silhouette to deter them. A fake owl or snake nearby often does the trick.

1 Tbsp (15 mL) canola oil
1 1/2 lbs (680 g) boneless, skinless chicken breast, cut into bite-size pieces
1 Tbsp (15 mL) minced ginger root

1 Tbsp (15 mL) canola oil
2 Tbsp (30 mL) rice vinegar
2 Tbsp (30 mL) teriyaki sauce
1 Tbsp (15 mL) honey
1 lb (454 g) fresh sweet cherries, pitted and halved
1 1/2 cups (375 mL) shredded carrot
1/2 cup (125 mL) chopped shallots
1/3 cup (75 mL) sliced almonds, toasted

12 x 10 inch (25 cm) flour tortillas
2 cups (500 mL) shredded lettuce

Heat first amount of canola oil in large skillet over medium-high. Add chicken and ginger and cook for 7 to 10 minutes, until chicken is done. Set aside.

In large bowl, whisk together second amount of canola oil, rice vinegar, teriyaki sauce and honey. Add chicken mixture, cherries, carrot, shallots and almonds and toss to combine.

Spoon mixture onto centre of each wrap, top with lettuce and serve.

Cherry

Easy Canadian Cherry Pie

Serves 6

There are two basic varieties of cherries available on the market: sour cherries and sweet cherries. Sweet cherries are more common and also more in demand. In Canada, the majority of sweet cherry growing is carried out in British Columbia, mostly in the Okanagan and Kootenay regions, which account for approximately 75 percent of all sweet cherries produced in this country. The majority of the remaining 25 percent is grown in Southern Ontario. Canada has worked to increase its production of sweet cherries in recent years to meet worldwide demand; however, the growing season is too short in most of the country for the fragile trees.

1/4 cup (60 mL) butter
1/4 cup (60 mL) brown sugar
1 tsp (5 mL) vanilla extract
3/4 cup (175 mL) flour
1/2 cup (125 mL) rolled oats
3 Tbsp (45 mL) water

2 cups (500 mL) pitted fresh sweet cherries
1/2 cup (125 mL) granulated sugar

Preheat oven to 400°F (200°C). In medium bowl, blend butter, brown sugar and vanilla, then add flour, rolled oats and water. Mix well. Press mixture into bottom and up sides of 9 inch (23 cm) pie dish.

In another medium bowl, mix together cherries and sugar. Spoon mixture into crust. Bake in oven for 12 to 15 minutes, until crust is golden brown.

Cherry Upside-down Cake

Serves 8

Upside-down cake has been made to the great delight of dessert enthusiasts for decades. The classic upside-down cake was made with pineapple, but the traditional fruit glaze topping shines best, in my opinion, when sweet Canadian cherries are used. This recipe calls for a round baking dish, but I have also put together some terrific Cherry Upside-down Cakes with the use of a cast-iron skillet. When using a cast-iron skillet, you almost need to dedicate a special skillet for this purpose, since any skillet previously used to sauté onions or fry a steak would be seasoned with flavours that may not go well with this cake. Whatever way you cook it, this dessert is always a real crowd-pleaser.

> 3 Tbsp (45 mL) butter, softened
> 6 Tbsp (90 mL) brown sugar
> 1 lb (454 g) fresh sweet cherries, pitted and halved
>
> 1 cup (250 mL) flour, packed
> 2/3 cup (150 mL) granulated sugar
> 2 tsp (10 mL) baking powder
> 1/2 tsp (2 mL) salt
> 1/8 tsp (0.5 mL) ground nutmeg
> 1/2 cup (125 mL) milk
> 1/3 cup (75 mL) butter, softened
> 1 tsp (5 mL) vanilla extract
> 1 egg
>
> 1/2 cup (125 mL) whipping cream
> 1 tsp (5 mL) granulated sugar

Preheat oven to 350°F (175°C). Add first amount of butter to greased 8 inch (20 cm) round baking pan, and melt in preheating oven. Add brown sugar and mix, then spread evenly over bottom of pan. Starting at outer edge, arrange cherries, cut side up, in single layer on bottom of pan; press lightly to make them stick. Set aside.

In large bowl, combine flour, first amount of granulated sugar, baking powder, salt and nutmeg; mix. Stir in milk, second amount of butter and vanilla. Beat in egg. Spread batter carefully over cherries. Bake for 35 to 40 minutes, until wooden pick inserted in centre comes out clean. Let cool in pan on wire rack for 5 minutes. Loosen edges with knife, then invert cake onto platter; wait for 1 minute or so before removing pan. Allow to cool for about 45 minutes.

(continued on next page)

134

Whip cream and second amount of granulated sugar together until stiff. Add a spoonful of whipped cream to each serving.

Pictured on page 124.

Cherry Cobbler

Serves 6 to 8

Although sour cherries tend not to be good eating fresh, they are absolutely perfect for many recipes. Some chefs prefer cooking with sour cherries because they hold together better, and their natural tartness mellows into a mild sweetness as they cook. With sour cherry demand on the increase in Canada, more and more commercial growers are getting in on the action. Currently Canada imports more sour cherries than it grows, but it is predicted that in coming years, our reliance on sour cherry imports will be lessened. Most sour cherries are grown in orchards; however, many backyards also play host to this terrific red fruit.

1 cup (250 mL) flour
3/4 cup (175 mL) granulated sugar
1 tsp (5 mL) baking powder
1/4 tsp (1 mL) salt
1/2 cup (125 mL) milk
3 Tbsp (45 mL) butter, melted

3 1/2 cups (875 mL) pitted fresh sour cherries

1 cup (250 mL) granulated sugar
1 Tbsp (15 mL) cornstarch
1 cup (250 mL) boiling water

Preheat oven to 350°F (175°C). In medium bowl, combine flour, first amount of sugar, baking powder and salt. Mix. Add milk and butter, and mix well to form dough.

Place cherries in 9 x 9 inch (23 x 23 cm) pan. Spread dough over cherries.

In a small bowl, combine second amount of sugar and cornstarch. Stir in boiling water. Pour cornstarch mixture over dough. Bake for 45 minutes. Serve warm.

Canadian Cherry Chutney

Makes 4 cups (1 L)

The production of sour cherries has really taken off in Canada in recent years, especially in Ontario and the Prairie Provinces. Although not as popular from a pick and eat standpoint, this variety of cherry is better suited to our northern climate and is more hardy than the sweet variety. Sour cherry growers have really fine-tuned their operations over the past 10 years. New strains of sour cherries have been developed that respond very well to our climate and growing season. Within the province of Ontario, the southern counties of Haldimand and Norfolk are the new bustling area for sour cherries and represent 25 percent of the production for the entire province.

> 1 lb (454 g) fresh sweet or sour cherries, pitted and halved
> 1 apple, peeled and chopped
> 1 large onion, diced
> 1 1/2 cups (375 mL) apple cider vinegar
> 1/4 cup (60 mL) granulated sugar
> 1/4 cup (60 mL) brown sugar
> 2 Tbsp (30 mL) minced ginger root
> 1 tsp (5 mL) salt
> 1/4 tsp (1 mL) ground nutmeg

Combine all 9 ingredients in large saucepan over medium-high. Bring to a simmer, stirring occasionally. Reduce heat to low. Cover and simmer for 45 minutes, stirring occasionally. Remove lid and continue to simmer for 20 to 30 more minutes, until desired consistency has been achieved. Refrigerate for 2 hours before serving with old-fashioned ham or wildfowl. It will keep well in a sealed container in the refrigerator for up to 3 months.

Cherry

Cherry Cranberry Martini

Serves 1

The delectable cherry is widely used in cooking and can be included in a variety of recipes. When selecting cherries in the store, there are a few tips for finding the best ones. Look for cherries that are not bruised, appear darker red than other cherries nearby and feel soft to the touch. Be sure also to check if it is sweet or tart cherries that you are selecting. The sweet cherries are great for eating fresh, but the tart ones may only be suitable for cooking. Cherries can be stored in the fridge for up to a week, and you should wash them only just prior to use.

    4 ice cubes
    3 Tbsp (45 mL) vodka
    2 Tbsp (30 mL) cherry juice
    2 Tbsp (30 mL) cranberry juice
    2 Tbsp (30 mL) pineapple juice
    1 fresh sweet cherry, pitted

Fill cocktail shaker with ice; add vodka, cherry juice, cranberry juice and pineapple juice. Shake vigorously and then strain into martini glass. Garnish with cherry and serve.

Pictured on page 69.

Chokecherry Apple Butter

Makes 8 cups (2 L)

The beloved chokecherry should have been doomed from the get go, with an inherent bitterness and sourness that is considered by most people unappetizing. Still, Canadian people have used the chokecherry extensively; somehow we have been able to get past the extremely sour taste and discover its many uses, including this Chokecherry Apple Butter. For a berry that is virtually inedible fresh, it makes terrific jams and jellies.

> 4 cups (1 L) fresh chokecherries
> 1 cup (250 mL) water
>
> 4 cups (1 L) applesauce
> 5 cups (1.25 L) granulated sugar
> 1/2 tsp (2 mL) almond extract

Place chokecherries in large saucepan and add water. Simmer on low for 10 to 12 minutes, until berries are soft. Put fruit through sieve or food mill to remove stones. Measure 2 cups (500 mL) chokecherry pulp back into saucepan.

Add applesauce. Bring to a boil, stirring often. Add sugar and continue to cook, stirring constantly, for 10 minutes, until it just begins to thicken. Add almond extract and mix. Remove from heat. Carefully ladle hot mixture into hot sterile jars, leaving 1/4 inch (6 mm) headspace. Wipe rims clean. Place hot metal lids on jars and screw on metal bands fingertip tight. Process in boiling water bath for 10 minutes. Remove jars and allow to cool. Check that jars have sealed properly. Store sealed jars in a cool, dark place.

 CHECKING JAR SEALS

Properly sealing jars of jelly, jam and other preserves protects against spoiling. After the jars are processed in a boiling water bath and have cooled, check the seals; the lids should be popped inwards. If you push on the top, it will not "click" or make a noise because the pressure has sucked the lid in. If the jar has not sealed properly, there may be too much headspace, or there may be a flaw with the jar; check for cracks. Any jars that are not properly sealed should be stored in the refrigerator and used first.

Traditional Canadian Chokecherry Jam

Makes 7 cups (1.75 L)

Foraging for chokecherries is something I did as a child. I can recall visiting my Aunt Betty in Arundel, Québec, and picking scads of chokecherries each summer. My aunt made wonderful chokecherry jam, and she always invited us to help her gather a supply of the tiny cherries large enough to make an enormous batch of her jam. It was easy to go visit her regardless of the reason, but certainly the scenic beauty of her property perched high on a hill overlooking that valley in Laurentian Mountains made chokecherry picking easy on the eyes.

4 cups (1 L) fresh chokecherries
1 cup (250 mL) water

1/2 cup (125 mL) lemon juice
1/4 tsp (1 mL) butter
1 x 2 oz (57 g) box pectin crystals
6 cups (1.5 L) granulated sugar

Crush chokecherries and place in large saucepan. Add water. Simmer over medium for 10 to 12 minutes, until berries are soft. Put fruit through sieve or food mill to remove stones. Measure 3 1/2 cups (875 mL) chokecherry pulp back into saucepan.

Add lemon juice and butter (butter helps to reduce foaming). Mix. Add pectin. Mix well. Bring to a boil over high, stirring constantly. Remove from heat. Add sugar and stir until all sugar is dissolved. Continue stirring. Return to heat. Bring to a full rolling boil and boil hard for 1 minute, stirring constantly. Remove from heat. Skim off any foam. Stir and skim for 5 minutes to cool slightly and to reduce floating fruit. Carefully ladle mixture into hot sterile jars, leaving 1/4 inch (6 mm) headspace. Wipe rims clean. Place hot metal lids on jars and screw on metal bands fingertip tight. Process in boiling water bath for 15 minutes. Remove jars and allow to cool. Check that jars have sealed properly. Store sealed jars in a cool, dark place.

Chokecherry Syrup

Makes 6 cups (1.5 L)

Chokecherries (*Prunus virginiana*) grow on small bushes across most of southern Canada. Their fruits range in colour from dark red to black and tend to grow in clumps or clusters. Like all *Prunus* species, chokecherries possess large stones. There are seven species of *Prunus* native to Canada, with the chokecherry being one of the most crucial naturally occurring fruits to aboriginals. Native Canadians used chokecherries for everything from making astringent and wine to dyes for their clothing. The chokecherry is about as Canadian as the beaver or the loon.

6 cups (1.5 L) fresh chokecherries
water

6 cups (1.5 L) granulated sugar

Place chokecherries in large saucepan and add just enough water to barely cover berries; cover pot. Bring to a boil over high; reduce heat and simmer for 5 to 7 minutes, until berries are soft. Mash berries, then transfer fruit to strainer set over bowl. Do not squeeze or press fruit.

Measure 3 cups (750 mL) juice back into saucepan. Bring to a boil. Add sugar and stir until it is dissolved. Return to a boil and boil for 15 minutes, stirring occasionally. Pour into hot sterile jars. Wipe rims clean. Place hot metal lids on jars and screw on metal bands fingertip tight. Store in a cool, dark place.

Summer Berry Meringue Dessert Nests (p. 158)

Bumbleberry Pie (p. 164)

Chokecherry Shrub Drink

Makes about 4 cups (1 L) concentrate

Shrubs, known as drinking vinegars, are primitive refreshing drinks from the colonial days before the invention of energy drinks or soft drinks. Before refrigerators and freezers were invented, one way to keep berries for longer periods of time was to preserve them in the form of shrub. It was often mixed with cool water to provide a pick-me-up during summer. An old-fashioned shrub has a flavour that is both sweet and sour, so it stimulates the appetite while quenching thirst. Try your chokecherry shrub with ice and club soda. It is a rare and unique drink.

3 cups (750 mL) fresh chokecherries (see Note)
1 cup (250 mL) white vinegar
1 cup (250 mL) water

sugar, according to amount of juice extracted

In large bowl, mash chokecherries. Combine vinegar and water and pour over berries (solution should cover berries). Let stand for 24 hours, stirring occasionally.

Transfer mixture to large saucepan, and slowly bring to a boil over medium-high. Boil for 5 minutes. Remove from heat. Strain mixture through jelly bag or cheesecloth. Do not squeeze bag. Return extracted juice to saucepan. For every 1 cup (250 mL) juice, add 1 cup (250 mL) sugar. Bring to a boil and boil for 2 minutes. Remove from heat and set aside to cool. Bottle and refrigerate for up to 6 months.

To Serve: Add 2 to 4 Tbsp (30 to 60 mL) chokecherry shrub over ice in tall glass. Top with water or club soda.

Note: Blueberries or saskatoon berries may be substituted for chokecherries in this recipe.

Chokecherry Liqueur

Makes 8 cups (2 L)

The chokecherry has been important in Canada for many years and is very popular in putting together a variety of alcoholic beverages including wine, sherry and liqueur. Chokecherry liqueur has been described as being similar in makeup to the Italian liqueur Limoncello, where sugar is added in the fermentation process, and grain alcohol is used. Fermenting chokecherry liqueur the long way is extremely labour intensive. I much prefer this chokecherry liqueur recipe; it is quick because there is no fermentation, and the vodka will more than do the job.

4 cups (1 L) fresh chokecherries
4 cups (750 mL) sugar
4 cups (750 mL) vodka

Wash two 4 cup (1 L) glass jars with a screw top that can be tightly closed. To each jar, add 1 cup (250 mL) chokecherries then 1 cup (250 mL) sugar. Repeat. Pour 2 cups (500 mL) vodka over mixture in each jar. Cover. Turn jars upside down, and flip once each day for 6 weeks. Strain contents, and you are ready to taste. Serve straight in a 2 oz (60 mL) liqueur glass or over ice.

Elderberry Corn Pancakes

Serves 6

Did you know that many of our Canadian flowers are not only nice to look at but also good to eat? Eating flowers will require some research and effort so as not to consume anything that might be toxic. For starters, try adding some peppery nasturtiums to a salad, or adding some fragrant rosewater to your baking. The rule of thumb for edible flowers is that the safe ones will taste similar to how they smell; if you like their fragrance, you will like their flavour too. Elderflowers are another edible flower. If you just can't wait for the berries to be ready, try using the flowers as a replacement in this recipe—just be sure to remove the bitter stalks first. You may be pleasantly surprised.

1 1/3 cups (325 mL) flour
1/2 cup (125 mL) yellow cornmeal
2 tsp (10 mL) baking powder
1 tsp (5 mL) baking soda
1/4 tsp (1 mL) salt

2 eggs
1 cup (250 mL) milk
2 Tbsp (30 mL) honey
1 Tbsp (15 mL) butter, melted
1 tsp (5 mL) vanilla extract

1 cup (250 mL) fresh (or frozen, thawed) elderberries

In large bowl, combine flour, cornmeal, baking powder, baking soda and salt. Mix.

In medium bowl, beat eggs; add milk, honey, butter and vanilla and whisk together. Make a well in centre of dry ingredients and pour wet ingredients into well. Whisk batter until smooth.

Preheat greased griddle to medium. Pour about 1/4 cup (60 mL) batter onto griddle. Sprinkle berries onto each pancake as soon as you pour it on griddle. When deep holes or bubbles appear in pancake, flip and cook other side until golden brown. Repeat for each pancake. Serve with syrup.

Mulberry Muffins

Makes 12 muffins

Although they grow quite prevalently in other parts of North America, especially in the southern United States, mulberries are extremely rare tree species in the Canadian wild. The three main species of mulberry found in Canada are the red, black and white mulberry. The red mulberry is in fact an endangered species in Ontario, which is ironic because these trees are extremely easy to grow. In recent years, the majority of mulberries grown in Canada have been hand-planted, and there is an initiative to continue hand-planting them. Mulberry trees may reach 10 metres in height and produce huge numbers of berries—perfect for enjoying in this muffin recipe.

> 1 cup (250 mL) self-rising flour
> 1 cup (250 mL) caster (or granulated) sugar
> 2 tsp (10 mL) ground cinnamon
> 1 tsp (5 mL) baking soda
>
> 1 egg
> 1 cup (250 mL) butter, melted
> 1 cup (250 mL) milk
> 1 cup (250 mL) fresh or frozen mulberries

Preheat oven to 350°F (175°C). In large bowl, combine flour, sugar, cinnamon and baking soda. Mix.

In medium bowl, whisk together egg, butter and milk. Add to flour mixture and stir just until combined. Fold in mulberries. Spoon batter into 12 greased muffin cups. Bake for 20 to 25 minutes, until wooden pick inserted in centre of muffin comes out clean.

Pictured on page 123.

Nagoonberry Squares

Serves 12

Foraging for berries in the wild is a terrific outdoor activity, with the added bonus that you just never know what sort of wildlife you might encounter. For Maggie Cruikshank Qingalik, while berry picking in Nunavik back in late summer 2012, there was a lot more than the typical animals roaming around the berry patch. Qingalik said the creature she saw appeared out of nowhere and sported long, shaggy hair, stood at least 3 metres (10 feet) tall and left huge footprints. The woman also claimed the animal, which she believes was a Sasquatch, did not appear vicious or aggressive and had little interest in her or her berry picking. And the most I ever see are deer—some people have all the luck!

1/4 cup (60 mL) butter
2/3 cup (150 mL) brown sugar
1 Tbsp (15 mL) vanilla extract
1 egg, beaten

1 cup (250 mL) flour
1 tsp (5 mL) baking powder
1/2 tsp (2 mL) salt
1/2 tsp (2 mL) ground cinnamon
1/2 cup (125 mL) fresh or frozen nagoonberries
1/2 cup (125 mL) chopped walnuts or almonds

Preheat oven to 350°F (175°C). In small saucepan over medium, melt butter. Remove from heat. Stir in brown sugar, vanilla and egg.

In medium bowl, combine flour, baking powder, salt and cinnamon. Add butter mixture and mix well. Gently fold in nagoonberries and nuts. Pour batter into 8 x 8 inch (20 x 20 cm) baking pan and bake for about 35 minutes, until lightly browned. Let cool before cutting into squares.

Salmonberry Crisp

Serves 6

The salmonberry was an extremely important species for west coast aboriginal peoples, who used all parts of the salmonberry plant, not just the tasty berry. During the early spring growing season, salmonberry sprouts were peeled apart and eaten raw with salmon meat or salmon roe, which is the origin of the name. Because of its ability to grow quickly and its intricate root system, the salmonberry bush is often planted along streams and rivers to prevent soil erosion in springtime. Salmonberries are related to blackberries and raspberries; like raspberries, they pull away from the receptacle when picked ripe for a hollow berry.

4 cups (1 L) fresh or frozen salmonberries
1/2 cup (125 mL) granulated sugar
1 Tbsp (15 mL) cornstarch
1 tsp (5 mL) vanilla extract

1/2 cup (125 mL) butter
1/2 cup (125 mL) rolled oats
1/2 cup (125 mL) flour
1/2 cup (125 mL) granulated sugar
1/2 cup (125 mL) slivered almonds
1/2 tsp (2 mL) salt

Preheat oven to 350°F (175°C). In large bowl, combine salmonberries, first amount of sugar, cornstarch and vanilla and mix well. Spread mixture in bottom of greased 9 x 9 inch (23 x 23 cm) baking pan.

Place 6 remaining ingredients in food processor and pulse until mixture is crumbly. Sprinkle over berry mixture. Bake for approximately 45 minutes, until top has browned slightly.

Soapberry Indian Ice Cream

Makes 6 cups (1.5 L)

Although they grow from Newfoundland to the Yukon and were heavily used by aboriginal peoples, soapberries are a rare item to most Canadians. Soapberry, or buffalo berry, as it is often called, is edible but bitter if harvested too early. Soapberries are best picked later in the season, after the plants have been exposed to at least one frost, when the taste of berry improves because the sugar content rises. They come in different colours; yellow soapberries tend to be sweeter. They are high in pectin and contain seeds that are easily consumed along with the fruit.

1 cup (250 mL) fresh soapberries
1 cup (250 mL) water
1/4 cup (60 mL) granulated sugar

Place soapberries and water into wide-topped ceramic, metal or glass mixing bowl (do not use a plastic bowl or utensils, and make sure that nothing is greasy, or the berries will not whip properly). Whip mixture with electric eggbeater or hand whisk until it reaches the consistency of beaten egg whites. Gradually add sugar, but not too fast, or the foam will "sink." Serve immediately.

Thimbleberry Dressing

Makes 2 cups (500 mL)

I will never forget the day I discovered the thimbleberry. I had taken my young daughters out for a bike ride near our Ontario home; the last hill before our house was too steep to peddle, so we decided to walk. I enjoy pointing out resident flora and fauna to my kids, and we came upon a patch of what I thought were wild raspberries, but which later turned out were thimbleberries. Once I had identified them correctly, we quickly returned with a small container to do some picking. In my opinion, the thimbleberry beats the wild raspberry hands down with its inherent sweetness and superior size. The thimbleberry's large size means your picking containers fill up quickly.

> 1 cup (250 mL) fresh thimbleberries
> 1/2 cup (125 mL) extra-virgin olive oil
> 1/4 cup (60 mL) apple cider vinegar
> 1 tsp (5 mL) granulated sugar
> 2 garlic cloves, minced
> 1 tsp (5 mL) salt

In large bowl, crush berries. Add 5 remaining ingredients. Whisk vigorously and drizzle over salad. It will keep well in a sealed container in the refrigerator for up to 3 months.

Bunchberry Raspberry Syrup

Makes 1 cup (250 mL)

Found growing across the boreal forests of Canada and particularly abundant in the hardwood forests of eastern Ontario and western Québec, bunchberry is also known as creeping dogwood. The berries are green during spring and turn a bright red later in summer as they approach ripeness. They typically grow together in a tight bunch—hence the name. The bunchberry is the only dogwood to grow mere inches off the ground; other dogwoods are more tree-like. Bunchberries are mild and flavourful, tasting of a hint of apple.

> 2 cups (500 mL) fresh or frozen bunchberries
> 2 cups (500 mL) fresh or frozen raspberries
> 1/4 cup (60 mL) water
> 1/4 cup (60 mL) honey

(continued on next page)

Combine all 4 ingredients in large saucepan and cook on medium-high for about 10 minutes, until fruit is soft. Remove from heat. Press berry mixture through fine-mesh sieve or colander to separate syrup from pulp. Serve over pancakes, French toast or waffles. It will keep well in a sealed container for up to 3 months in the refrigerator.

Crabapple Jelly
Makes 6 cups (1.5 L)

Although technically not a berry, crabapples are the only apple native to North America and are often included along with mention of wild berries of Canada. The crabapple has some terrific culinary uses, with crabapple jelly being one of the most popular. If you have ever mistakenly taken a bite of a crabapple thinking it was a regular apple, you were in for quite a shock. The sourness of these fruits is instantaneous; however, when some sugar is added, a charming flavour emerges. Crabapple jelly is simple to make; crabapples contain so much natural pectin that there is no need to add any extra.

8 cups (2 L) stemmed and quartered fresh crabapples
water, to cover

4 cups (1 L) granulated sugar

Place crabapple quarters in large stainless steel or other non-reactive saucepan. Add just enough water so apples are covered but not floating. Bring to a boil over high. Reduce heat to medium and simmer for 10 to 15 minutes. Crabapples should soften and change colour. Line strainer with 2 or 3 layers of cheesecloth and place over large bowl. Pour crabapple mixture into strainer; cover with edges of cheesecloth. Let stand for 30 minutes, or until liquid measures 4 cups (1 L).

Discard pulp, and pour juice back into saucepan. Bring to a simmer and cook for 10 minutes. Skim off any foam. Add sugar and stir until completely dissolved. Continue cooking at a low boil until temperature reaches 220°F (105°C). Remove from heat. Skim off any foam. Carefully ladle hot mixture into hot sterile jars, leaving 1/4 inch (6 mm) headspace. Wipe rims clean. Place hot metal lids on jars and screw on metal bands fingertip tight. Process in boiling water bath for 10 minutes. Remove jars and allow to cool. Check that jars have sealed properly. Store sealed jars in a cool, dark place.

Highbush Cranberry Blueberry Pemmican

Makes 6 cups (1.5 L)

Pemmican is a traditional source of protein that could be stored for long periods of time, often flavoured with berries. It was typically made using big-game animals such as moose and bison. The meat was cut into long, thin strips and slow-dried over a low fire. The dried strips were then pounded into tiny pieces and mixed with melted animal fat to hold the concoction together in a solid mass. In many cases, dried berries such as blueberries or saskatoons, or occasionally other available berries such as bush cranberries, were added to the mixture. The pemmican was then moulded into a ball and wrapped in a rawhide bag for long-term storage. This modern version of pemmican, with added spices and sugar and no melted animal fat, is slightly less labour intensive and hugely more palatable.

3 Tbsp (45 mL) butter
3 Tbsp (45 mL) brown sugar
1/4 tsp (1 mL) ground ginger
1/4 tsp (1 mL) ground cloves
1/4 tsp (1 mL) ground cinnamon

2 cups (500 mL) fresh (or frozen, thawed) highbush cranberries
2 cups (500 mL) fresh (or frozen, thawed) wild blueberries

4 cups (1 L) finely chopped beef jerky
1/2 cup (125 mL) chopped walnuts (optional)
1/2 cup (125 mL) sunflower seeds (optional)

In large saucepan over medium, combine butter, brown sugar, ginger, cloves and cinnamon. Heat, stirring, until butter is melted.

In medium bowl, combine both berries and mash. Add to saucepan. Bring mixture to a simmer, stirring constantly, and allow to simmer for about 5 minutes. Remove from heat and set aside to cool.

Preheat oven to 175°F (80°C). Add jerky, walnuts and sunflower seeds to cooled berry mixture. Spread mixture evenly on greased rimmed baking sheet. Cook in oven for about 6 hours, until completely dried out. Allow to cool. Tear into snack-sized pieces.

Baked French Toast
with Nuts and Berries

Serves 6 to 8

Down home, earthy recipes such as this one are always best enjoyed in the great outdoors. While camping with my family or even when visiting the hunt camp in the off season, I will often prepare Baked French Toast with Nuts and Berries. Everyone in our clan loves French toast, and berries and nuts add an interesting and irresistable element to that classic. Of course, topping it off with some Canadian maple syrup is always a must. All French toast connoisseurs have a supply of natural syrup ready at any given time.

6 thick slices of bread
4 eggs
1/2 cup (125 mL) brown sugar
1/4 tsp (1 mL) ground cinnamon
1/4 tsp (1 mL) nutmeg
2 cups (500 mL) milk
1 tsp (5 mL) vanilla extract

4 cups (1 L) fresh or frozen mixed berries
1 cup (250 mL) chopped pecans
1/4 cup (60 mL) brown sugar
1/4 cup (60 mL) butter, melted

Arrange bread in single layer on bottom of greased 9 x 13 inch (23 x 33 cm) baking pan. In large bowl, combine eggs, first amount of brown sugar, cinnamon and nutmeg and whisk until well blended; whisk in milk and vanilla. Pour evenly over bread. Cover and refrigerate for at least 6 hours or overnight.

Preheat oven to 400°F (200°C). Take bread mixture out of refrigerator. Spread berries evenly over bread. In small bowl, combine pecans, second amount of brown sugar and butter; sprinkle evenly over berries. Bake for 35 to 40 minutes, until bread is puffed in centre and fruit is bubbling.

Bumbleberry Muffins

Makes 12 muffins

I love creating bumbleberry frozen packets. During the peak of berry foraging season, I will mix three or four different berry types together and freeze them in 1 cup (250 mL) portions for use later in the season. The berry types are up to you; you can make each packet the same, or create a variety of mixes. Because blackberries, raspberries and blueberries all come into ripeness around the same time in my region, many of my bumbleberry bags are composed of those three varieties. Later, when winter has descended on the Canadian landscape, you can hunt through your freezer for a bumbleberry bag to brighten your day.

2 cups (500 mL) flour
1 1/4 cups (300 mL) granulated sugar
2 tsp (10 mL) baking powder
1/2 tsp (2 mL) salt

2 eggs
1/2 cup (125 mL) vegetable oil
1/2 cup (125 mL) milk
2 cups (500 mL) fresh or frozen mixed berries

Preheat oven to 350°F (175°C). In large bowl, combine flour, sugar, baking powder and salt. Mix.

In medium bowl, whisk together eggs, oil and milk. Add to flour mixture and mix just until combined. Gently fold in berries. Spoon batter into 12 greased muffin cups. Bake for 25 to 30 minutes, until wooden pick inserted in centre of muffin comes out clean.

Bumbleberry

Fruity Salsa with Cinnamon Bites

Serves 10 as a snack

This recipe is a terrific snack you can make in advance. The Fruity Salsa can be put together and refrigerated in an airtight container, and the Cinnamon Bites can be stored in a resealable plastic bag. I like to make this recipe and bring it with me to the cottage for the weekend to serve to guests as a snack on Saturday evening. Several snacks like this can be made in advance and enjoyed during times of leisure when you don't necessarily want to be slaving in the kitchen. Your friends and family will appreciate your forward thinking.

2 kiwi fruits, peeled and diced
2 apples, diced
1 1/2 cups (375 mL) fresh blueberries
1 1/2 cups (375 mL) fresh raspberries
2 cups (500 mL) diced fresh strawberries
3 Tbsp (45 mL) strawberry jam (see Note)

10 x 10 inch (25 cm) flour tortillas
3 Tbsp (45 mL) butter, melted

1 1/2 tsp (7 mL) granulated sugar
1/2 tsp (2 mL) ground cinnamon

In large bowl, combine first 6 ingredients. Mix. Cover and refrigerate for at least 1 hour.

Preheat oven to 350°F (175°C). Brush 1 side of each tortilla with butter. Cut into wedges and arrange on baking sheets.

Combine sugar and cinnamon, and sprinkle over wedges. Bake for 8 to 10 minutes, until crispy. Allow to cool and place on serving platter. Remove chilled Fruity Salsa from fridge and serve with Cinnamon Bites.

Note: If you don't want to use store-bought jam, you can make your own; Easy No-cook Strawberry Freezer Jam, page 107, will work well in this recipe.

Berry Fruit Leather

Makes 1 baking sheet of fruit leather

Here is another recipe that allows you to mix and match berries to your liking. You might call fruit leather "vegetarian beef jerky." It is an intriguing treat that, although especially popular with children, is appreciated by people of all ages. You can make this recipe in advance and, after the fruit leather has been cut into strips, each one can be individually wrapped and brought away on a picnic lunch as a snack. It is certainly not run of the mill.

4 cups (1 L) crushed fresh berries (use mixed or all one kind)
1/2 cup (125 mL) granulated sugar

2 cups (500 mL) applesauce

Stir crushed berries and sugar together in medium saucepan over medium until sugar has dissolved. Put mixture through food mill or sieve to remove any stems or seeds.

Add applesauce and mix well. Pour onto greased rimmed baking sheet and spread mixture to an even thickness. Place in food dehydrator, or in oven at 175°F (80°C), for 5 to 6 hours, until firm to the touch and dry enough to peel off. Let cool completely. Use scissors to cut fruit leather into strips, and store strips in airtight container or bag.

Crunchy Bumbleberry Cheese Balls

Serves 6 to 8 as an appetizer

Pairing your berry recipe with pistachios is as tasty as it is nutritious. Not only are pistachios one of my favourite nuts, but like berries, they are also extremely healthy and energy rich. They are an excellent source of vitamin E and B vitamins such as riboflavin and niacin. They boast high levels of such minerals as copper and manganese, as well as a fatty acid called oleic acid, which is a proven antioxidant. Regular consumption of pistachios has been known to lower the LDL, or bad cholesterol, in our bodies. This is one terrific nut.

> 1 cup (250 mL) chopped pistachios, *divided*
> 8 oz (250 g) cream cheese
> 4 oz (125 g) goat cheese
> 3/4 cup (175 mL) dried mixed berries, chopped
> 1 tsp (5 mL) lemon juice
> 1/2 tsp (2 mL) salt
> 1/2 tsp (2 mL) pepper

In large bowl, combine 3/4 cup (175 mL) pistachios, both cheeses, dried berries, lemon juice, salt and pepper. Cream together ingredients using electric mixer. Refrigerate for 2 hours.

Scoop tablespoon-size balls of cheese mixture and roll in remaining 1/4 cup (60 mL) pistachios to coat. Serve as an appetizer.

Summer Berry Meringue Dessert Nests

Serves 6 to 8

Summertime and berries go hand in hand. Everything from the foraging to
the preparation and preserving, and using them in recipes ranging from simple
sauces and pies to fancy, more extravagant desserts are also a summertime
tradition. These Summer Berry Meringue Dessert Nests, for example, are
a creation that almost needs to be carried out with freshly picked berries.
Each bite is a taste of summer.

    5 egg whites
    1/4 tsp (1 mL) cream of tartar
    1 cup (250 mL) granulated sugar
    1/4 tsp (1 mL) ground cinnamon

    3 cups (750 mL) fresh mixed berries
    2 Tbsp (30 mL) granulated sugar

    1 cup (250 mL) whipping cream
    1/2 cup (125 mL) icing sugar
    1 Tbsp (15 mL) orange liqueur

Preheat oven to 250°F (120°C). Line baking sheet with parchment paper.
In large bowl, beat egg whites with cream of tartar until foamy. Slowly
add first amount of granulated sugar, and then add cinnamon, continuing
to beat until meringue is glossy and forms stiff peaks. Spread about 1/2 cup
(125 mL) meringue into 3 inch (7.5 cm) circle on parchment paper. Shape
rim around edge, forming nest-like shape. Continue making nests until all
meringue is used. Bake for about 1 1/2 hours, until firm to the touch. Turn
off oven, but leave meringue nests in warm oven for another hour. Remove
from oven and allow to cool completely.

Meanwhile, in medium bowl, combine berries with second amount
of granulated sugar. Set aside for at least 2 hours.

Combine whipping cream with icing sugar. Whip until stiff peaks form. Fold
in orange liqueur and 1 to 2 tsp (5 to 10 mL) of berry juices, and refrigerate
for 1 to 2 hours before serving. To serve, place an individual meringue nest
on each plate, top with some whipped cream mixture, and add a spoonful
or two of berries on top.

Pictured on page 141.

Bumbleberry

Photo Index to the Fruits

Blackberry

Blueberry

Bunchberry

Cherry

Chokecherry

Crabapple

Cranberry (Highbush)

Cranberry (Lowbush)

Currant (Red)

Elderberry

Photo Index to the Fruits

Gooseberry

Huckleberry

Mulberry

Nagoonberry

Raspberry

Salmonberry

Saskatoon

Soapberry

Strawberry

Thimbleberry

Bumbleberry Crumble

Serves 8

Some confusion exists around the term "bumbleberry." Bumbleberries are actually not a species of berry at all—they don't exist. The word bumbleberry simply describes a random mix of berries, generally used in pies and other dishes. Bumbleberry mixtures are usually composed of several types of berries that happen to be in season at the time, that you may stumble—or bumble—upon while out foraging. Depending on the mixture, variances in taste will result.

3 cups (750 mL) fresh or frozen mixed berries
2 medium apples, peeled, cored and sliced
1/4 cup (60 mL) brown sugar
1/3 cup (75 mL) apple juice
1 Tbsp (15 mL) lemon juice
1 Tbsp (15 mL) cornstarch

1 1/4 cups (300 mL) rolled oats
1/4 cup (60 mL) butter, softened
3 Tbsp (45 mL) brown sugar
2 Tbsp (30 mL) flour
1/2 tsp (2 mL) ground cinnamon
1/8 tsp (0.5 mL) ground nutmeg

Preheat oven to 350°F (175°C). In large bowl, combine berries, apples, first amount of brown sugar, apple juice, lemon juice and cornstarch. Mix well. Spoon into 8 x 8 inch (20 x 20 cm) baking pan.

In medium bowl, combine rolled oats, butter, second amount of brown sugar, flour, cinnamon and nutmeg. Mix well. Sprinkle over fruit mixture. Bake for 40 minutes, or until golden brown and bubbly. Serve warm or at room temperature.

Fruit Pizza

Serves 8

When deciding what to do with berries, certain obvious recipes come to mind; however, pizza is not usually one of them. Trying new things doesn't always come as second nature to me, so it took some coaxing to try Fruit Pizza. I am sure glad that I did, because this pizza boasts pizazz and is nothing like you've ever eaten before. More of a dessert than a main course, its real niche may not yet be determined. All I know is, once you slice it up, you'll have hungry folks from all around running to the table.

1 1/4 cups (300 mL) flour
2/3 cup (150 mL) butter, softened
1/2 cup (125 mL) granulated sugar

1 x 8 oz (250 g) block cream cheese, softened
1/3 cup (75 mL) icing sugar
1 tsp (5 mL) vanilla extract

1 x 10 oz (284 mL) can mandarin orange segments, drained
1/2 cup (125 mL) fresh blueberries
24 fresh strawberries, halved lengthwise
8 kiwi fruits, sliced

1/4 cup (60 mL) apricot (or peach) jam
1 Tbsp (15 mL) hot water

Preheat oven to 350°F (175°C). Mix flour, butter and granulated sugar in medium bowl until mixture forms a ball. Press firmly in ungreased 12 inch (30 cm) pizza pan, forming rim around edge. Bake for about 12 minutes, until golden. Cool.

Beat cream cheese, icing sugar and vanilla in small bowl until smooth. Spread evenly over crust.

Arrange all 4 types of fruit in attractive pattern over cream cheese mixture.

Combine jam and hot water in small cup. Brush over fruit. Chill before serving.

Slow Cooker Bumbleberry Cobbler

Serves 8

Slow cookers are one of the greatest cooking inventions ever. One of their best attributes is their ability to take a moderately tough piece of protein and turn it into the most delectable meal you've ever tasted. Another is that the low cooking temperature means the food will rarely stick to the dish, and you will never burn a meal. The ability to start your meal before heading off to work in the morning, and then to arrive home with it piping hot waiting for you, makes life easier. There are even several recipes, like this Slow Cooker Bumbleberry Cobbler, for dessert. The flexibility and ease of use make the slow cooker a fabulous tool that every cook should become familiar with.

1 cup (250 mL) flour
3 Tbsp (45 mL) granulated sugar
1 tsp (5 mL) baking powder
1/4 tsp (1 mL) ground cinnamon

1 egg
1/4 cup (60 mL) milk
2 Tbsp (30 mL) vegetable oil

1/4 cup (60 mL) flour
1 cup (250 mL) granulated sugar
1/8 tsp (0.5 mL) salt
4 cups (1 L) fresh or frozen mixed berries
1 1/2 tsp (7 mL) lemon juice

In large bowl, combine first amounts of flour and sugar with baking powder and cinnamon. Mix.

In small bowl, whisk together egg, milk and oil. Add to flour mixture and stir just until moistened. Spread over bottom of slow cooker dish.

In medium bowl, combine second amounts of flour and sugar with salt. Add berries and lemon juice. Stir to coat berries with flour mixture. Evenly distribute berry mixture over batter in slow cooker. Cover and cook on Low for 2 to 2 1/2 hours, until batter is cooked through. Serve warm.

Bumbleberry Pie

Serves 8

1 3/4 cups (425 mL) flour
1 Tbsp (15 mL) brown sugar
3/4 tsp (4 mL) salt
1/4 tsp (1 mL) baking powder
1/3 lb (150 g) cold vegetable shortening

3/4 cup (175 mL) cold water, approximately
2 tsp (10 mL) white vinegar

1 cup (250 mL) granulated sugar
1 Tbsp (15 mL) lemon zest
3 Tbsp (45 mL) corn starch
1/2 tsp (2 mL) salt
4 cups (1 L) fresh mixed berries

1 large egg, beaten

In large bowl, combine flour, brown sugar, salt and baking powder. Cut in shortening until mixture resembles coarse crumbs.

In small bowl, combine water and vinegar. Slowly add to flour mixture, stirring with a fork, until mixture starts to come together. You may not need to use all liquid. Do not over mix. Turn out onto work surface. Shape into 2 slightly flattened discs, one slightly larger than other. Wrap each with plastic wrap. Chill for 1 hour before rolling. Roll out larger dough portion on lightly floured surface to about 1/8 inch (3 mm) thick. Line 9 inch (23 cm) pie dish.

Preheat oven to 350°F (175°C). In large bowl, combine sugar, lemon zest, corn starch and second amount of salt. Mix well. Add berries and mix to coat. Spread in pie shell. Roll out smaller portion of dough on lightly floured surface to about 1/8 inch (3 mm) thick. Cut out several small vents with a small cookie cutter. Dampen edge of pastry shell in pie dish and cover with remaining pastry. Trim and crimp decorative edge to seal.

Brush crust with egg. Bake on bottom rack oven for 45 to 55 minutes, until crust is golden and berries are tender.

Pictured on page 142.

Bumbleberry Sorbet

Serves 4

The Prairie Fruit Growers Association is a non-profit organization representing all the fruit crop growers in Manitoba, including all the berry operations. The association is represented by elected members who work tirelessly on food safety and environmental farm plan initiatives to ensure that public standards in Manitoba-grown fruit are met not only for taste but for safety as well. The PFGA website is a great place to look for "U-pick" farms in your area and also to find some tips on foraging wild berries in the province of Manitoba— precisely what you need to enjoy your mixed berry sorbet.

1 1/4 cups (300 mL) water
1/4 cup (60 mL) granulated sugar

2 cups (500 mL) fresh or frozen mixed berries
3 Tbsp (45 mL) water

2 Tbsp (30 mL) orange liqueur
2 tsp (10 mL) lemon juice

2 egg whites

In small saucepan over medium, combine first amount of water and sugar. Heat, stirring, for about 2 minutes, until sugar is dissolved. Increase heat to medium-high and bring to a boil. Boil gently for 10 minutes. Set aside to cool.

In medium saucepan over medium, combine berries and second amount of water. Cook for about 3 minutes, stirring occasionally, until berries are softened and broken up. Press through fine-mesh sieve into medium bowl. Discard seeds. Add berries to sugar water.

Add orange liqueur and lemon juice. Mix well. Spread evenly in ungreased 1 1/2 quart (1.5 L) shallow baking dish. Freeze for about 2 hours, until almost firm.

In separate medium bowl, beat egg whites until soft peaks form. Scrape frozen berry mixture into egg whites. Fold until no white streaks remain. Spread evenly in same baking dish. Freeze for about 2 hours, until firm. Scrape mixture into blender or food processor. Process until smooth. Spread evenly in same baking dish. Cover. Freeze for about 2 hours, until firm. Serve frozen.

White Chocolate Berry Soufflé

Serves 6

Soufflé calls for a bit of skill, but it is a dish that will satisfy as it impresses. Although this recipe can be put together quite at any given time, I like to prepare White Chocolate Berry Soufflé for a special occasion. There is just something about the puffy goodness of a soufflé, combined with the rich smoothness of white chocolate, that is irresistible. Try experimenting with your berry mixture. Different combinations will provide a varied flavour palate. My favourite combination is half raspberry and half strawberry, but the choice is up to you.

2 cups (500 mL) fresh (or frozen, thawed) mixed berries
1/2 cup (125 mL) granulated sugar
2 Tbsp (30 mL) orange liqueur

1/2 tsp (2 mL) butter
2 Tbsp (30 mL) granulated sugar

2 Tbsp (30 mL) butter
2 Tbsp (30 mL) flour
1/2 cup (125 mL) milk
3 x 1 oz (28 g) white chocolate baking squares, chopped

2 egg yolks, fork-beaten
3 Tbsp (45 mL) granulated sugar

4 egg whites, room temperature

1 Tbsp (15 mL) icing sugar

Combine berries, first amount of sugar and liqueur in medium frying pan over medium. Heat and stir until sugar is dissolved. Increase heat to medium-high. Cook for about 5 minutes, stirring occasionally, until liquid is absorbed and mixture is slightly thickened. Set aside to cool.

(continued on next page)

Preheat oven to 375°F (190°C). Grease six 1/2 cup (125 mL) ramekins with first amount of butter. Sprinkle each ramekin with 1 tsp (5 mL) sugar. Tilt ramekins to coat bottom and side of each one. Gently tap to remove excess sugar. Place ramekins on baking sheet. Divide and spoon berry mixture into each one.

Melt second amount of butter in small saucepan over medium. Add flour. Heat and stir for 1 minute. Add milk. Stir for 2 to 3 minutes, until boiling and thickened. Add chocolate. Heat and stir until chocolate is almost melted. Do not overheat. Remove from heat and stir until smooth.

In small bowl, combine egg yolks, third amount of sugar and 2 Tbsp (30 mL) chocolate mixture. Stir into remaining chocolate mixture. Heat and stir on medium-low until sugar is dissolved. Transfer to large bowl.

In medium bowl, beat egg whites until stiff peaks form. Fold about 1/3 of egg whites into chocolate mixture until almost combined. Fold in remaining egg whites until just combined. Divide and carefully spoon mixture over berries in ramekins. Bake for about 20 minutes, without opening oven door, until puffed and set.

Dust with icing sugar. Serve immediately.

Bumbleberry Pudding Cakes

Serves 6

Before attempting to forage your own wild berries, it is important that you do your homework. Research the types of berries you're targeting, and also the region in which you will be picking. Familiarize yourself with any berries unsuitable for human consumption that grow in your area. You would hate to be filling your pails with poisonous berries, jeopardizing the health of you and your family. Search the internet for berry picking tips, and chat with locals in the area where you are planning on foraging. There is a wealth of information out there to enrich your picking endeavours.

3 cups (750 mL) fresh (or frozen, thawed) mixed berries

2 eggs
1/4 cup (60 mL) granulated sugar
1 tsp (5 mL) vanilla extract
1 cup (250 mL) milk
1/2 cup (125 mL) flour
1/2 tsp (2 mL) baking powder
1/8 tsp (0.5 mL) salt

Preheat oven to 400°F (200°C). Lightly coat six 6 ounce (170 mL) ramekins with cooking spray. Arrange on baking sheet. Divide berries among ramekins. Set aside.

In medium bowl, combine eggs, sugar and vanilla; whisk until light and frothy. Whisk in milk until combined. Add flour, baking powder and salt and whisk until smooth. Pour batter over berries. Bake for about 20 minutes, until golden brown. Serve warm.

Triple Berry Jam

Makes 12 cups (3 L)

My Triple Berry Jam typically includes strawberries, wild raspberries and wild blueberries but has also been made with other of the great alternatives we have. Even switching up wild berries for the cultivated variety can alter the flavour of your jam. Wild blueberries have a distinctly different, tangier flavour than cultivated blueberries, and wild raspberries tend to produce a sweeter, more full-bodied jam than do cultivated raspberries. Given the difficulty of actually picking enough wild strawberries, it is much easier just to make use of cultivated strawberries when they are in season.

> **4 cups (1 L) fresh wild blueberries**
> **2 1/2 cups (625 mL) fresh wild red raspberries**
> **2 1/2 cups (625 mL) fresh strawberries**
> **1/4 cup (60 mL) lemon juice**
> **2 x 2 oz (57 g) boxes pectin crystals**
>
> **11 cups (2.75 L) granulated sugar**

Crush fruit and place in large saucepan. Add lemon juice and mix well. Stir in pectin. Bring to a full rolling boil over high, stirring constantly.

Stir in sugar. Return to a full rolling boil for 1 minute, stirring constantly. Remove from heat. Skim off any foam. Carefully ladle hot mixture into hot sterile jars, leaving 1/4 inch (6 mm) headspace. Wipe rims clean. Place hot metal lids on jars and screw on metal bands fingertip tight. Process in boiling water bath for 15 minutes. Remove jars and allow to cool. Check that jars have sealed properly. Store sealed jars in a cool, dark place.

Berry Vegetable Smoothie

Serves 2

In recent years, the popularity of the vegetable smoothie has taken flight, and now with the introduction of berries into this healthy beverage, you've got the best of both worlds. The nutrient-dense combination of spinach, which is high in dietary potassium, coconut, which provides fibre, and vitamin C, along with a healthy serving of mixed berries, which is full of antioxidants and other essential nutrients, is unbeatable as a healthy combination. Besides that, the Berry Vegetable Smoothie is just plain refreshing!

> 1 cup (250 mL) frozen mixed berries
> 1 1/2 cups (375 mL) fresh spinach leaves, lightly packed
> 2 cups (500 mL) coconut milk
> 1 Tbsp (15 mL) coconut oil
> 1 cup (250 mL) ice cubes

Combine all 5 ingredients in blender and blend on low for 30 seconds. Increase to high and blend for 1 minute, or until smooth throughout. Divide into 2 large glasses.

Any Berry Milkshake

Serves 2

The versatility of this recipe is part of its appeal; you will find your own personal favourite combinations of berries as you go along. Mixing berries in different ways is fun, and it adds a unique and varying flavour each time you make an Any Berry Milkshake. I usually construct this drink with whatever wild berries I have on hand, but any berry, as the name implies, will do the trick.

> 1 1/2 cups (375 mL) ice cubes
> 2/3 cup (150 mL) milk
> 2/3 cup (150 mL) frozen mixed berries
> 2 Tbsp (30 mL) granulated sugar
> 1/4 tsp (1 mL) vanilla extract

Fill blender with ice cubes. Add milk, berries, sugar and vanilla. Blend until smooth. Divide into 2 glasses and serve with a straw.

Pictured on page 123.

Recipe Index

Index   **175**

About the Author

Growing up in a family-owned hotel in the Laurentian Mountains of rural Québec, Jeff was introduced to the outdoors and great cooking at a very young age, falling in love with both instantly. Over the years, he has made the great outdoors a focal point for his life's work. Jeff has a degree in environmental management as well as fish and wildlife biology. He is an award-winning member of the Outdoor Writers of Canada and has contributed to several Canadian and American publications over the years. He currently writes for Newfoundland's popular *Outdoor Sportsman* magazine and *Outdoor Canada* magazine, and has a regular column in *Bounder Magazine*. His first book, *Weird Facts about Fishing*, was released in 2010, and he writes a popular blog, *The Outdoors Guy*, for the *Ottawa Sun*.

Jeff has travelled to each and every province from coast to coast—hunting, fishing, camping and enjoying the fruits of his labour. He describes himself as the consummate conservationist and family-man, and describes his cooking as down to earth, simple, and about as Canadian as you can get. Jeff spent a lot of time at his uncle's famous steakhouse in the mountains of Québec, and picked up copious down-home tips along the way. He brings with him an in-depth knowledge of nature and conservation and a genuine love and passion for the outdoors—from the field to the table.